D0581861

The Continuing Quest

LARGE-SCALE OCEAN SCIENCE FOR THE FUTURE

The Continuing Quest

LARGE-SCALE OCEAN SCIENCE FOR THE FUTURE

Prepared by the

Post-IDOE Planning Steering Committee
Ocean Sciences Board
Assembly of Mathematical and Physical Sciences
National Research Council

NATIONAL ACADEMY OF SCIENCES WASHINGTON, D.C. 1979

NOTICE: The project that is the subject of this report was approved by the Governing Board of the National Research Council, whose members are drawn from the councils of the National Academy of Sciences, the National Academy of Engineering, and the Institute of Medicine. The members of the Committee responsible for the report were chosen for their special competences and with regard for appropriate balance.

This report has been reviewed by a group other than the authors according to procedures approved by a Report Review Committee consisting of members of the National Academy of Sciences, the National Academy of Engineering, and the Institute of Medicine.

Ship details shown on the cover, the title page, and pages xii, xviii, 5, 24, 31, 42, and 83, are from a map prepared in 1775 by Thomas Jeffreys, Geographer to His Majesty, showing the flotilla track followed from Vera Cruz, Mexico, to Havana, Cuba, which avoided the trade winds and hurricanes. The map is owned by Harris B. Stewart, Jr., and hangs in the Atlantic Oceanographic and Meteorological Laboratories of the National Oceanic and Atmospheric Administration, Miami, Florida.

Library of Congress Cataloging in Publication Data

National Research Council. Post-IDOE Planning Steering Committee.
The continuing quest.

1. Oceanographic research–United States. 2. Submarine geology–Research–United States. I. Title.
GC58.N34 1979 551.4′6′0072073 78-20892
ISBN 0-309-02798-5

Available from:

Office of Publications
National Academy of Sciences
2101 Constitution Avenue, N.W.
Washington, D.C. 20418

Printed in the United States of America

Post-IDOE Planning Study Steering Committee

WARREN S. WOOSTER, University of Washington, Seattle, Washington, *Chairman*

JOHN V. BYRNE, Oregon State University, Corvallis, Oregon

REUBEN LASKER, National Marine Fisheries Service, La Jolla, California

FOSTER H. MIDDLETON, University of Rhode Island, Kingston, Rhode Island

BRIAN J. ROTHSCHILD, National Marine Fisheries Service, Washington, D.C.

DEREK W. SPENCER, Woods Hole Oceanographic Institution, Woods Hole, Massachusetts

FERRIS WEBSTER, Woods Hole Oceanographic Institution, Woods Hole, Massachusetts

Ocean Sciences Board

*OSB Chairman, January 1976–June 1978; now, Assistant Administrator for Research and Development, National Oceanic and Atmospheric Administration, Rockville, Maryland.

†Now, Dean, School of Oceanography, Oregon State University, Corvallis, Oregon.

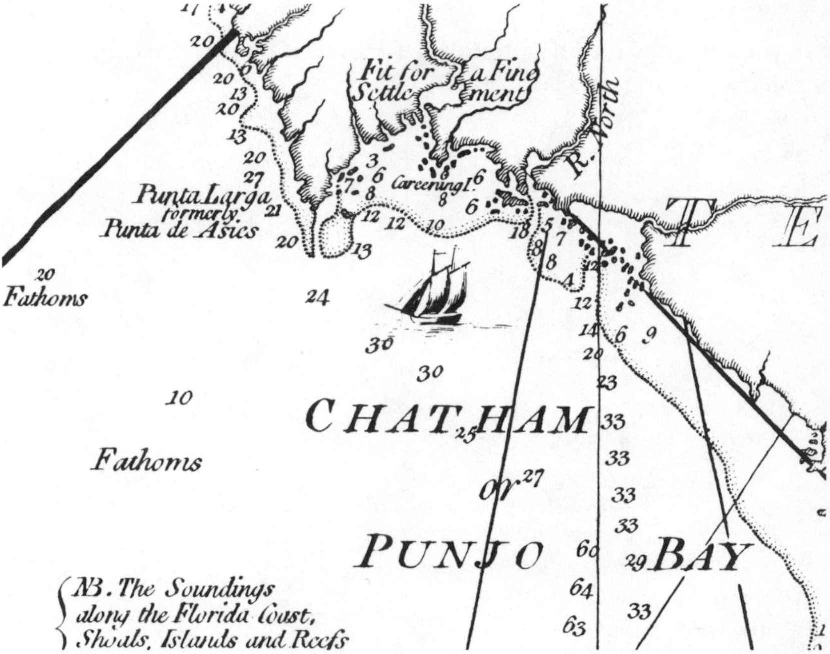

Fit for a Fine Settlement

R. North

Punta Larga formerly Punta de Asics

Careening I.

T E

20 Fathoms

10 Fathoms

CHATHAM or PUNJO BAY

NB. The Soundings along the Florida Coast, Shoals, Islands and Reefs

Contents

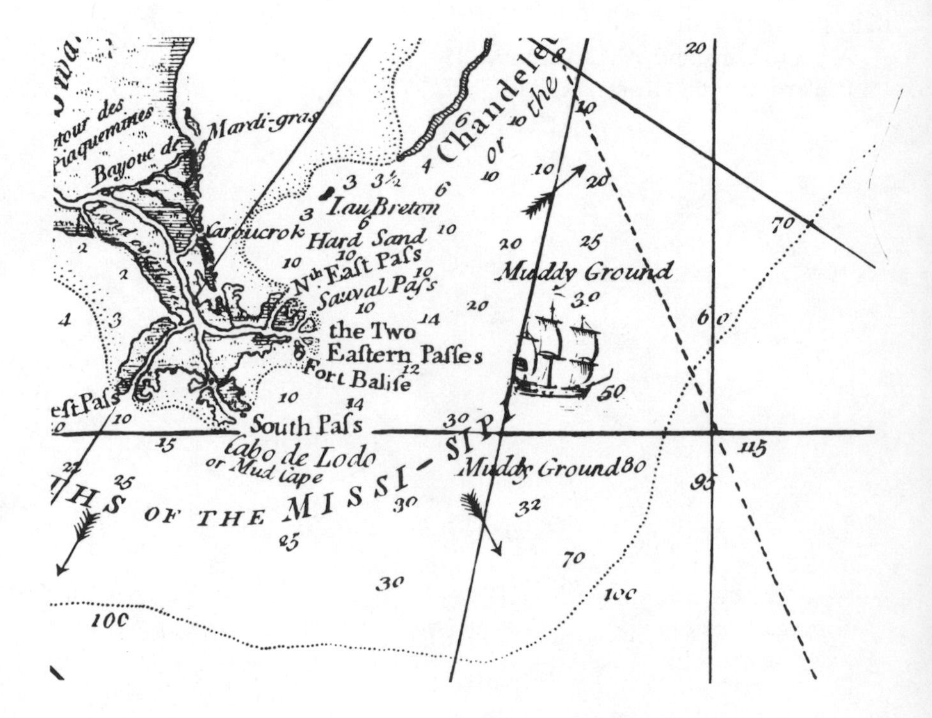

tour des Piaquemines
Bayoue de
Mardi-gras
Chandeleur
or the
I. au Breton
Hard Sand
Nth East Pass
Sauval Pass
the Two Eastern Passes
Fort Balise
Muddy Ground
South Pass
Cabo de Lodo
or Mud Cape
Muddy Ground
THS OF THE MISSI-SIP

Introduction

BACKGROUND

Since its inception as an organized field of scientific inquiry more than a hundred years ago, oceanography has evolved from a subject of growing individual inquiry to a large, complex, multi-institutional, and expensive enterprise. The evolution has not been steady, proceeding by fits and starts. A major turning point for oceanography in the United States occurred at the conclusion of World War II when the federal government initiated a program of substantial and continued support for ocean investigations. This program reflected not only the broader national support for scientific inquiry concerning the oceans but also recognition of the many ways in which oceanic knowledge could be applied to the major national needs for food, energy, mineral resources, and security.

Consequences of the postwar growth were the establishment of new oceanographic institutions and the expansion of existing ones, the increased training of marine scientists, and the construction of laboratories and research vessels, in short, the development of a major U.S. marine-science capability.

During the 1950's and 1960's, much of oceanographic research was conducted by individuals or small groups of scientists working within single disciplines. This research gradually revealed the magnitude of the complexities and interactions that characterize the ocean system. It became evident that the size and complexity of some problems required investigations that were joint efforts involving the scientists and facilities of several institutions and disciplines.

A means for such joint action was provided by the establishment, in 1969, of the International Decade of Ocean Exploration (IDOE), a National Science Foundation (NSF) program devoted to large-scale, cooperative oceanic research concerned not only with increased scientific knowledge and understanding but also with the expectation that its findings could result in the more effective utilization of the ocean and its resources. The conceptual basis for the program and many of its characteristic features were developed by a group of scientists and engineers organized by the National Academy of Sciences and the National Academy of Engineering. Their views were expressed in the 1969 National Academy of Sciences publication entitled *An Oceanic Quest.*

The IDOE has made possible an attack on problems that resisted the efforts of individual scientists and has fostered cooperation and the pooling of expertise and facilities among institutions and disciplines to an extent hitherto impossible. Some major scientific accomplishments of the IDOE are described in the present report.

In the middle of the present decade, scientists and administrators began to consider continuation of the large-scale, cooperative approach to ocean investigations beyond the time span of the IDOE. The effectiveness of the IDOE approach had been demonstrated, and it was clear that many scientific problems of the future would require even greater efforts for their solution.

POST-IDOE PLANNING

In February 1974, H. Guyford Stever, Director of the National Science Foundation, requested the National Advisory Committee on Oceans and Atmosphere (NACOA) to review the IDOE and to make recommendations regarding its future. NACOA, after a detailed study of the program, issued a report in August 1975 entitled *The International Decade of Ocean Exploration: A Mid-Term Review.* In this report, NACOA endorsed continuation of NSF support for long-term, multidisciplinary, multi-institutional ocean studies and proposed that action be taken to define goals and guidelines for the programs that should succeed the IDOE.

The NACOA recommendations were reviewed and supported by the NSF IDOE Advisory Panel, which had provided broad policy guidance to the IDOE since its beginning in 1970, and by the *ad hoc* IDOE Advisory Panel of the Ocean Sciences Board and the Marine Board of the National Research Council. In early 1977, the NSF initiated the organization of four workshops to be held that spring in each of the oceanographic disciplines: physics, biology, chemistry, and geology and geophysics. Specialists listed in the NAS *U.S. Directory of Marine Scientists* were invited to submit their views on important scientific opportunities and issues that should be considered in the post-1980 program. Some 400 members of the oceanographic community responded during the spring and summer of 1977. These replies were important contributions to the disciplinary workshops and to the reports that followed.

In February 1977, the NSF asked that the National Academy of Sciences and the National Academy of Engineering continue to provide advice and guidance on the nature of programs to follow the IDOE. Specifically, advice was requested on participants for the disciplinary workshops, on reviews of discussion papers, and on recommendations for new areas of research, particularly those requiring multidisciplinary approaches. In addition, the Foundation expressed its hope that the Academies would assume responsibility for organizing, conducting, and preparing a study based on deliberations from these workshops and on other sources that would result in specific recommendations for a successor to the IDOE. That study and the preparation of this report were sponsored by funds provided by the Foundation. To carry out this work, the Ocean Sciences Board established a Steering Committee with representation from the Marine Board and the Ocean Policy Committee.

In selecting members of the Steering Committee, a balance was struck between experience in the IDOE program and independence from that program. Two members were deeply involved in major IDOE projects, two members had no direct involvement but had previously directed laboratories that received support from IDOE, and three members had neither received support nor held administrative posts related to the IDOE.

The four disciplinary workshops were convened by the NSF, in cooperation with the University of Rhode Island's Center for Ocean Management Studies, in the spring of 1977. Participants were asked to identify those problems in their own disciplines that marked the most promising directions for research. These problems were considered primarily from the perspectives of the current and anticipated state of knowledge in the disciplines and the availability of the technology necessary to attempt to solve these problems. Participants also examined the rationale for project support; processes of project initiation, review, and administration; international aspects; the interaction of marine research with that on adjacent land; and the need and methods for transfer of the results to application.

In addition, the Steering Committee organized a fifth workshop to discuss in greater detail issues of program initiation and management, the international aspects of the proposed program, and the interdisciplinary research opportunities not identified in the other workshops. The reports of these five workshops and lists of participants and respondents to the preworkshop inquiry were brought together in a single volume, *Ocean Research in the*

1980's, Recommendations from a Series of Workshops on Promising Opportunities in Large Scale Oceanographic Research, published by the Center for Ocean Management Studies, University of Rhode Island, August 1977.

The Steering Committee organized a final summary workshop held at the Battelle Northwest Research Center in Seattle, Washington, September 7–14, 1977. Some 80 individuals, including marine and social scientists, laboratory administrators, federal agency officials, industrial managers, and foreign scientists, took part in discussions about research needs and opportunities for large-scale, long-term oceanographic studies during the 1980's. No separate report of this workshop was prepared, but its discussions and recommendations had an important influence on the present report.

CONTENT OF THE REPORT

In preparing this report, the Steering Committee has drawn heavily on the above-described actions. The credibility of our conclusions and recommendations is derived from the broad participation of the ocean-science and engineering community.

This report is a response to a request from the National Science Foundation and is primarily addressed to that agency. As an element of the national ocean research program, the proposed extension of the IDOE will also be of interest to other agencies and offices of the Executive and Legislative Branches of the federal government. The report contains ideas contributed by many scientists, engineers, and other ocean users, who are another segment of the audience. Finally, the public continues to be intrigued by oceanic endeavors and will find matters of interest in the report.

The report's conclusions and recommendations are listed in Chapter 1 and are cross-referenced to the pages on which they are discussed in more detail.

Chapter 2 contains a review of the origins of the IDOE and a description of some of the program's major scientific accomplishments. We have not attempted an exhaustive critique of the program. This was done by NACOA in its midterm review. As that report points out, "The entire IDOE program has been subjected to repeated scientific review by NSF advisory panels and by the National Academy of Sciences and the National Academy of Engineering." We have nonetheless drawn upon the experience of the past, and have taken criticisms of the IDOE into account, in considering the character of the proposed future program.

Chapter 3 reviews the interactions between the proposed program of fundamental research and the eventual application of its findings to the uses of the ocean and its resources and to other marine-related programs of society. The argument is advanced that although these applications may be far in the future, fundamental research is required to provide the underlying understanding on which shorter-term application-oriented investigations can be based.

Scientific problems that might be studied under the proposed program are discussed in Chapter 4. Some central questions in each of the oceanographic disciplines are listed, and an interdisciplinary framework is presented that suggests some of the interactions among these questions and the investigations that might be proposed to answer them. The sample projects are intended only to be illustrative and cannot be assigned priorities. The development of ideas differs from example to example. As the program is implemented, these or other projects, both within and among disciplines, will be discussed by interested scientists and will be elaborated into proposals amenable to scientific and budgetary evaluation. Since this chapter is likely to be of greatest interest to scientists, no attempt has been made to simplify the specialized language occasionally employed.

Chapter 5 discusses management and administrative aspects of the future program as well as its requirements for personnel, facilities, and equipment. Finally, the estimated costs of the program are presented.

While recognizing that there are other substantial oceanographic proposals that will complete for funding, we are convinced that a post-IDOE program of large-scale cooperative research is essential to the continuing quest for oceanic knowledge and understanding. It would support and extend other important ocean research programs, both within the NSF and in other agencies that support such research. This report does not propose specific scientific investigations but rather a mechanism whereby appropriate research can be supported. Specific investigations must arise from the scientists who would carry them out and must be subjected to the usual processes of review and evaluation on which effective academic research is successfully based.

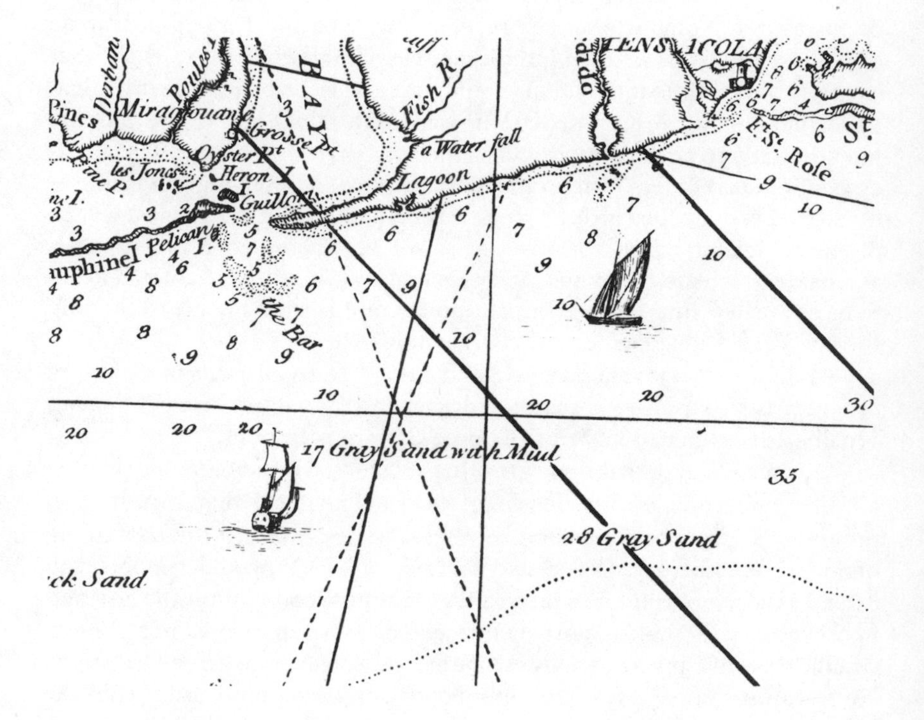
BAY
Fish R.
a Water fall
Lagoon
Miragouane
Grosse Pt
Oyster Pt
les Jones
Heron I.
Guillori
Pelican I.
the Bar
17 Gray Sand with Mud
28 Gray Sand
St. Rose
ck Sand

1 Principal Conclusions and Recommendations

PROPOSAL FOR A CONTINUING PROGRAM

1. A program of cooperative academic ocean research should follow and evolve from the International Decade of Ocean Exploration (IDOE). It should be sponsored by the National Science Foundation (NSF) as a major component of its overall efforts in fundamental oceanic research and should support and extend other important ocean research programs, both within the NSF and in other agencies that support such research. (For background discussions on this recommendation, see all chapters, especially Chapter 2, pages 12–24, and Chapter 4.)

2. The objective of this program should be to seek the comprehensive knowledge of ocean characteristics and their changes and the profound understanding of oceanic processes required for more effective utilization of the ocean and its resources, for protection of the marine environment, and for the prediction of natural events such as weather and climate. (See all chapters, especially Chapters 3 and 4, and page 61.)

3. The program is justified on the grounds of its potential for generating new scientific knowledge of the ocean and its interactions with the atmosphere, land, and seabed and for providing the essential conceptual framework on which the relationship between human activity and the marine environment should be based. (See all chapters, especially Chapters 3 and 4.)

4. The priority of the proposed program relative to other recently proposed marine research programs of comparable magnitude should be exam-

ined by the Ocean Sciences Board and other competent bodies concerned with the overall balance of the national oceanographic program. (See page xiv.)

5. Post-IDOE projects should be distinguished by features such as the following:

(a) A high probability of producing a significant increase in fundamental knowledge and understanding;

(b) Participation of a number of principal investigators, normally from several institutions;

(c) A significant degree of cooperation among scientists from different disciplines and countries;

(d) Funding level and project duration significantly larger than normally required by individual investigators;

(e) An identifiable relevance to the societal objectives of the program. (See pages 61–63.)

6. Projects may be suitable for inclusion if features such as the following apply:

(a) The size, complexity, and duration of the field programs impose special requirements for coordination;

(b) Instruments, equipment, and facilities are large and complex, and their sharing among institutions is appropriate;

(c) Large and expensive equipment must be developed;

(d) Field programs and modeling efforts must be integrated;

(e) The magnitude and complexity of the project requires planning and management procedures like those of the IDOE. (See page 62.)

7. The program should include pilot projects and other projects of intermediate size, larger than those normally supported by the Oceanography Section of NSF but smaller than major projects of the IDOE. Such projects may continue for a year or two and then either terminate or develop into major projects. (See page 63.)

8. An increase in funding beyond the level of the IDOE is proposed because of the following considerations:

(a) Increased national priority for ocean research arising from the growing potential for conflict among uses and users of the ocean and the urgency for a stronger scientific basis for ocean-policy decisions;

(b) The addition of projects of intermediate scale;

(c) The need to upgrade and replace elements of the academic fleet;

(d) Replacement and development of major items of equipment;

(e) Additional costs of conducting research in foreign waters. (See pages 80–83.)

9. In view of these considerations, we recommend that the proposed program be funded at a level (in 1981 dollars) increasing from $44.9 million

in 1981 to $57.9 million in 1990 (ship operating expenses included). (See pages 80–83.)

10. Support for ocean research at the level of the individual investigator should grow at a rate comparable with that of the proposed program. (See page 81.)

PROGRAM ORGANIZATION AND MANAGEMENT

11. A variety of approaches to project development should be encouraged. It is particularly important that the process be open to all scientists who are qualified and interested in participation. (See pages 63–64.)

12. For major projects of long duration, widespread participation in their development should be encouraged through early notification of planning meetings. Depending on the nature of the proposed research, scientists and engineers from a variety of institutions and representatives of government agencies should be included. (See pages 63–64.)

13. Initial screening of research topics should be performed by a body with broad interests, for example, by an advisory body for the NSF Division of Ocean Sciences as a whole. (See page 64.)

14. The process of peer review should continue to emphasize scientific quality and significance and the qualifications of investigators. Review panels should seek opportunities to combine projects or elements thereof where there is significant overlap. (See page 64.)

15. Responsibilities for elements of the future program should be allocated among NSF program managers on a flexible basis, with projects concerned with related scientific, logistic, and operational problems grouped under a common program manager. (See pages 64–65.)

16. At the operating level, projects should continue to be managed under arrangements developed by the participating scientists. These arrangements should be highly responsive to scientific needs and able to accommodate new ideas as they arise. (See pages 65–66.)

17. Projects should be structured into logical work and time segments, and the review schedule should take advantage of natural plateaus in development. Project duration should be made clear by the early establishment of well-defined termination points. (See page 66.)

COOPERATIVE ARRANGEMENTS

18. Interinstitutional cooperation in the development and implementation of projects as practiced during the IDOE should be a central attribute of the future program. (See pages 69–72.)

19. The post-IDOE program should be an identifiable entity within the NSF Division of Ocean Sciences. The NSF should develop an effective arrangement for coordination of the several ocean-related research programs within the Directorate for Astronomical, Atmospheric, Earth, and Ocean Sciences. (See pages 70–72.)

20. Mission-oriented agencies with responsibilities relating to the ocean should participate in the planning and implementation of relevant projects of fundamental research under the future program and should support the academic components of the projects, both directly and by making available ships and other support facilities. (See page 72.)

21. The post-IDOE program should foster cooperation with scientists of other countries, particularly where this will contribute to the achievement of scientific objectives. International cooperation should be developed on as informal a basis as possible, yet be consistent with the formal requirements for governmental involvement where access to foreign zones is required. The NSF should continue to monitor international arrangements and should encourage and support those organizations that successfully promote cooperative scientific activities. (See pages 66–68.)

22. The NSF should take the lead in negotiating suitable arrangements for the funding of marine research-related technical assistance with the Department of State, NOAA (Sea Grant), and other appropriate federal agencies. (See page 68.)

SUPPORT REQUIREMENTS

23. To meet the critical personnel requirements of the future program, academic institutions involved must continue their efforts to identify and develop potential scientific leaders. The assignment and support of talented project managers are also essential to the success of the proposed program. Special attention should be given to the stable funding of experienced research support groups whose services could be valuable to successive projects. Provision should be continued in project budgets for the support of participating graduate students. (See pages 72–73.)

24. The implementation of a national plan for maintaining an effective academic research fleet and of a schedule for funding the upgrading and replacement of existing vessels when necessary should be pursued vigorously. The NSF, together with NOAA and the Navy Department, should explore the possibilities for increased use of government vessels in the proposed program. (See pages 73–75.)

25. Existing high-technology capabilities should be made available to academic research, for example on specific government or academic re-

search vessels where their use can be shared. Project budgets should provide for acquiring state-of-the-art instruments where they will be much more effective than those currently available. (See pages 76–77.)

26. The engineering capabilities of universities, government, and industry should be employed in the development, testing, and construction of the major equipment needed for the proposed program. Specifications should arise from expressed scientific needs, ultimate scientist users should be involved in all phases of the development, and the products of these enterprises should be available to scientists in a variety of appropriate projects. Where major facilities and equipment are of value to both academic and government research, their development should be supported by the agencies concerned. (See pages 76–78.)

27. Efforts of NSF, NOAA, and academic research institutions to ensure prompt and full reporting and storing of data resulting from the IDOE and its successor program should continue and be expanded to include such data manipulation and analysis as are required to facilitate utilization of the stored information. An appropriate mechanism should be established to foster the necessary interaction among producers, storers, and users of oceanographic data. See pages 78–79.)

28. Research results from the IDOE and its successor should be disseminated widely to a multidisciplinary audience including potential users and the public. Findings of potential applied interest should be brought to the attention of appropriate mission-oriented agencies. Improved ways to accelerate the transfer and application of marine research findings should be sought. (See pages 79–80.)

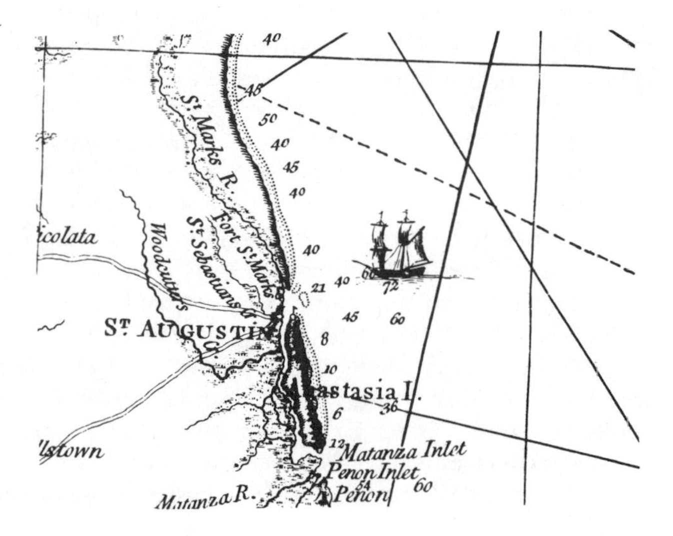

2 The International Decade of Ocean Exploration

HISTORICAL BACKGROUND

The IDOE was a watershed in the history of ocean research. By providing the structure and resources for large-scale, long-term coordinated projects, the program gave a powerful impetus to the transformation of marine science from a descriptive effort to one increasingly driven by experimental and theoretical concerns.

The descriptive phase of oceanography that extended into the early 1960's was constrained in its potential by conceptual and instrumental limitations and was restricted to a small number of research institutions.

Three kinds of cooperative oceanographic research were carried out in the first decades after the Second World War. The first involved large-scale cooperative investigations using several ships to make common measurements of features relating to particular oceanic variables. Examples include the cooperative studies of the Kuroshio, and the NORPAC, EASTROPAC, and EQUAPAC expeditions in the Pacific. In the Atlantic, similar efforts included environmental surveys for the Northwest Atlantic Fisheries Commission. The oceanographic program of the International Geophysical Year (1957–1959) involved 70 ships from 35 nations that took part in projects on geophysics, tides, deep-water circulation, and other problems related to the earth's climate and weather. Some 46 ships from 13 countries participated in the International Indian Ocean Expedition (1959–1965), which included important biological investigations as well as comprehensive studies of ocean circulation and of the geology and geophysics of the ocean basin.

A second approach entailed smaller expeditions to analyze and observe conditions in specific geographical locations or to make studies of particular marine phenomena. These included the multiship studies of the Gulf Stream, expeditions to study the equatorial circulation in the central and eastern Pacific, and expeditions to study the boundary currents, upwelling, and related biological phenomena off both coasts of South America.

The third utilized repeated sampling of a particular location or region to establish continuous time series at fixed points. Examples included the CALCOFI cruises off the west coast of the United States, the El Niño program of repeated sections at quarterly intervals off northern South and Central America, and the studies of the Gulf Stream.

The International Indian Ocean Expedition of the early 1960's was by certain criteria a forerunner of the IDOE program of the 1970's. Like the IDOE, the U.S. component of the program sought to organize the resources of the major oceanographic institutions in an international program to increase our knowledge concerning a largely unexplored part of the global ocean. But, as in most of the efforts preceding it, the scientists pursued their research as a loose confederation of individuals working on their own problems.

The mid-1960's saw an important shift in emphasis in marine activities. In the United States, the 1966 enactment of the Marine Resources and Engineering Development Act and creation of the National Sea Grant College Program reflected the growing concern for man's use and protection of the environment. That year, the United Nations General Assembly asked the Secretary General to survey the marine science and technology activities both of member States and of intergovernmental and nongovernmental international organizations and to compile proposals to bring about the most effective arrangements for an expanded program of international cooperation.

In March 1968, President Johnson announced endorsement of the concept of an International Decade of Ocean Exploration when he stated:

> The task of exploring the ocean's depth for its potential wealth—food, minerals, resources—is as vast as the seas themselves. No one nation can undertake that task alone. As we have learned from prior ventures in ocean exploration, cooperation is the only answer.
>
> I have instructed the Secretary of State to consult with other nations on the steps that could be taken to launch an historic and unprecedented adventure—an International Decade of Ocean Exploration for the 1970's.

Two months later the Intergovernmental Oceanographic Commission (IOC) adopted a formal recommendation supporting "the concept of an expanded, accelerated, long-term and sustained program of exploration of the oceans and their resources including the international programs, planned and coordinated on a world-wide basis." Further support came from the United Nations General Assembly, which in its Resolution 2467 (XXIII) of January 1969 endorsed "the concept of an International Decade of Ocean Exploration to be undertaken within the framework of a long-term programme of research and exploration. . . ."

At its Sixth Session in September 1969 the Intergovernmental Oceanographic Commission defined the purpose of the expanded program to be: "To increase knowledge of the ocean, its contents and the contents of its sub-soil, and its interfaces with the land, the atmosphere, and the ocean floor and to improve understanding of processes operating in or affecting the marine environment, with the goal of enhanced utilization of the ocean and its resources for the benefit of mankind. . . ."

The U.S. National Council on Marine Resources and Engineering Development then invited the National Academy of Sciences and the National Academy of Engineering to prepare detailed recommendations for the United States' contribution to the Decade. Experienced scientists and engineers from the academic, industrial, and governmental communities examined the full range of questions related to this effort. Attention focused on priorities among the scientific and engineering goals, the capabilities necessary to realize them, and the products and benefits to mankind anticipated from implementation of the Decade idea. In May 1969, the Academies jointly reported their conclusions in a report, *An Oceanic Quest, The International Decade of Ocean Exploration.*

The guiding premise of the International Decade concept was that sustained international planning and coordination would target on the most promising geographic areas and lines of scientific inquiry, set priorities, and agree on the sharing and distribution of effort. The results of this work would be published freely and promptly for the benefit of everyone. There was to be strong insistence on standardized data collection and dissemination, expanded activity by a large number of nations, and greater coordination among the international organizations concerned with the oceans. In short, the Decade was to be a period of "intensified collaborative planning among nations and expansion of exploration capabilities by individual nations, followed by execution of national and international programs of oceanic research and resource exploration so as to assemble a far more comprehensive knowledge of the sea in a reasonably short time."

As a major part of the Nixon Administration's program in marine science, Vice-President Agnew announced on October 19, 1969, the initial U.S. plans for participation in the International Decade of Ocean Exploration, and several weeks later, in his capacity as Chairman of the National Council on Marine Resources and Engineering Development, he assigned responsibility for the planning, management, and funding of United States

IDOE activities to the National Science Foundation. At this time, the following goals were set forth for the U.S. program:

• Preserve the ocean environment by accelerating scientific observations of the natural state of the ocean and its interactions with the coastal margin—to provide a basis for (a) assessing and predicting man-induced and natural modifications of the character of the oceans; (b) identifying damaging or irreversible effects of waste disposal at sea; and (c) comprehending the interaction of various levels of marine life to permit steps to prevent depletion or extinction of valuable species as a result of man's activities;

• Improve environmental forecasting to help reduce hazards to life and property and permit more efficient use of marine resources—by improving physical and mathematical models of the ocean and atmosphere which will provide the basis for increased accuracy, timeliness, and geographic precision of environmental forecasts;

• Expand seabed assessment activities to permit better management—domestically and internationally—of marine mineral exploration and exploitation by acquiring needed knowledge of seabed topography, structure, physical and dynamic properties, and resource potential, and to assist industry in planning more detailed investigations;

• Develop an ocean monitoring system to facilitate prediction of oceanographic and atmospheric conditions—through design and deployment of oceanographic data buoys and other remote sensing platforms;

• Improve worldwide data exchange through modernizing and standardizing national and international marine data collection, processing, and distribution; and

• Accelerate Decade planning to increase opportunities for international sharing of responsibilities and costs for ocean exploration, and to assure better use of limited exploration capabilities.

Shortly after receiving this charge, the National Science Foundation set up the Office for the International Decade of Ocean Exploration and began to define the U.S. program. In the first year of the Decade's existence, three areas were chosen for priority attention: (1) environmental quality, (2) environmental forecasting, and (3) seabed assessment. In 1971, living resources was added as a fourth program area.

CHARACTERISTICS AND ACCOMPLISHMENTS OF THE IDOE

One measure of the IDOE's impact on U.S. ocean research is its dollar contribution to the total national program (Table 1). Throughout its eight-year existence, it has funded annually $14 million to nearly $20 million of research, which constituted from about 11 to about 16 percent of the total federal oceanographic research support.

TABLE 1 Budgets for the IDOE, Federally Supported Oceanographic Research and the Total Federal Ocean Program 1971–1978 Estimated by Fiscal Year[a]

Fiscal Year	Total Federal Ocean Program ($ Millions)	Oceanographic Research ($ Millions)	IDOE[b] ($ Millions)	IDOE Percentage of Oceanographic Research (%)
1971	522.5	101.5	15.0[c]	14.7
1972	626.2	119.4	19.7[c]	16.4
1973	631.1	108.2	16.9[c]	15.6
1974	666.9	116.5	14.0	12.0
1975	782.5	124.1	14.9	12.0
1976	802.9	129.3	15.5	11.9
1977	943.5	149.8	17.2	11.5
1978	1,068.7	161.7	18.9	11.7
1979	1,139.0	174.9	19.7	11.3

[a]Sources: *Marine Science Affairs*, 1971; *The Federal Ocean Program*, 1972–1975; *The Federal Ocean Program–Budget Summary Fiscal Years 1975-1977* (March 1976); and Committee on Atmosphere and Oceans of the Federal Coordinating Council for Science, Engineering, and Technology.
[b]National Science Foundation, Office for the International Decade of Ocean Exploration, March 1978.
[c]Include $1.8 million, $2.3 million, and $2.8 million in ship support for each of these years; ship support in subsequent years is from the NSF Office for Oceanographic Facilities and Support.

Scientific research projects in the programs have sought to contribute to the 1969 goals enumerated above. Specifically, the Environmental Quality program has sponsored research aimed at understanding, assessing, and predicting the effects of man-made chemicals on the marine environment. Projects have included study of the effects of pollutants on marine environments, the ways in which pollutants get into the marine environment, and the worldwide distribution of the geochemical features of deep-ocean waters.

In Environmental Forecasting, investigators have sought to expand the scientific basis for improved environmental prediction. To do this, they have looked at historical climate changes, the influence of the oceans on the atmosphere, and the role played by ocean circulation in shaping weather and climate.

Seabed Assessment projects have been directed toward understanding the natural processes that have produced seabed metal and hydrocarbon deposits. Studies have been made of geological conditions that produce large accumulations of gas and oil on the continental margins; of the processes at midoceanic ridges and deep trenches and how they relate to the production of metal deposits; and of the processes at the deep-ocean floor that lead to manganese nodule formation.

Research in the Living Resources program has studied the ecological conditions needed to sustain marine life. Projects have included studies of upwelled, nutrient-rich waters in coastal areas of high fish productivity and of seagrass ecosystems.

Table 2 inventories some general characteristics of IDOE projects. These fall into four size classes:

I. Four projects, $21.5 million to $30.0 million, annual $2.1 million to $2.9 million, 11–16 institutions, 53 percent of total IDOE expenditures.

II. Six projects, $6.3 million to $12.0 million, annual $0.9 million to $1.5 million, 3–10 institutions, 29 percent of total IDOE expenditures.

III. Six projects, $4.0 million to $5.0 million, annual $0.7 million to $1.1 million, 5–9 institutions, 15 percent of total IDOE expenditures.

TABLE 2 Some Characteristics of Major IDOE Projects

Class	Name	Cost[a]	Rate[b]	Life[c]	Institutions[d]
I	NORPAX	30.0[e]	2.1	14	12
I	MODE/POLYMODE	23.0	2.9	8	16
I	GEOSECS	22.5	2.5	9	11
I	CUEA	21.5	2.7	8	15
II	Manganese Nodules	12.0	1.1	11	10
II	Nazca Plate	9.0	1.5	6	3
II	CEPEX	9.0	1.1	8	5
II	CLIMAP	8.4	0.9	9	7
II	ISOS	7.5	1.2	6	7
II	Continental Margins	6.3	1.3	5	7
III	SES	5.0	0.7	7	8
III	Metallogenesis	4.8	0.8	6	6
III	Pollutant Transfer	4.5	0.9	5	9
III	SEAREX	4.5	0.9	5	6
III	Biological Effects Laboratory	4.3	1.1	4	8
III	PRIMA	4.0	0.8	5	5
IV	Pollutant Baselines	2.6	1.3	2	17
IV	CENOP	2.5	0.8	3	11
IV	Mid-Atlantic Ridge	1.5	0.5	3	4

[a] In millions of dollars, total for project.
[b] Annual spending rate, in millions of dollars per year.
[c] Estimated duration in years.
[d] Number of participating institutions.
[e] Including about $15 million from ONR.

IV. Three projects, $1.5 million to $2.6 million, annual $0.5 million to $1.3 million, 4–17 institutions, 4 percent of total IDOE expenditures.

In contrast to IDOE projects, most projects of individual investigators from single institutions cost approximately $100,000 or less, and these projects usually continue for a year or two. Most major IDOE projects have been relatively expensive and long-lived.

The following section gives information on some of the scientific accomplishments of IDOE projects.* In a program of the magnitude of the IDOE, it is practicable to identify only some of the highlights in each field. Although these projects were administered in one or another of the application categories by which the IDOE Office has been organized, they usually focused on fundamental scientific problems, and their accomplishments can be evaluated within the framework of the traditional oceanographic disciplines—physics, biology, chemistry, and geology and geophysics. This is also true for interdisciplinary projects whose contributions to various disciplines are noted below.

PHYSICAL OCEANOGRAPHY

Physical oceanographic studies during the IDOE centered on the motion, composition, and structure of the fluid in the world ocean and on its interaction with the atmosphere and with its boundaries. Emphasis was on work directed at an understanding of the physical processes controlling such phenomena as ocean currents, upwelling, midocean turbulence, and atmospheric coupling. These experiments contrast with large-scale expeditions of earlier years that were concerned primarily with a more geographical description of the physical state of the ocean.

Analytical and theoretical models were developed in combination with the field work. The IDOE programs also provided an opportunity for significant advances in techniques for measurement and analysis.

The principal IDOE studies in physical oceanography were in the North Atlantic (MODE/POLYMODE), North Pacific (NORPAX), Southern Ocean (ISOS), and eastern boundary currents of the Pacific and the North Atlantic

*Detailed documentation on U.S. IDOE projects, including bibliographies of scientific reports and publications, is contained in a series of six progress reports prepared by the U.S. Department of Commerce, National Oceanic and Atmospheric Administration, Environmental Data Service, under contract to the National Science Foundation. The most recent of this ongoing series, published in October 1977, is entitled *International Decade of Ocean Exploration, Progress Report Volume 6, April 1976 to April 1977.*

(CUEA). Each of these projects had special objectives, but there are certain themes common to their accomplishments. These are summarized below.

VARIABILITY OF OCEAN PROCESSES

A predominant low-frequency variability has been identified by MODE/POLYMODE in the interior of the ocean. This variability is energetic and persistent, and the kinetic energy of the fluctuations is always greater than that of the local component of the mean circulation.

The dynamics of low-frequency fluctuations of the ocean–atmosphere system have been illuminated by NORPAX. In CUEA, time scales of upwelling processes have been documented and related to variables defining the ecosystem. The three-dimensional character of upwelling has been defined. Alongshore propagation of coherent current fluctuations in the upwelling region occurs over distances of 700 km. An important achievement of the ISOS project is the identification and description of bands of water in the Antarctic Circumpolar Current system and rings in the Polar Frontal Zone.

Mesoscale eddies in the interior of the ocean represent a special case of ocean variability. MODE/POLYMODE results demonstrate that these eddies are found in all oceans; at present their time and spatial scales are being defined. In view of this, it is not surprising that mesoscale eddies figure importantly in the results of both ISOS and NORPAX. Estimates of spatial scales of mesoscale turbulence have been obtained in ISOS, and evidence suggests that the Antarctic Circumpolar Current is an eddy-generating area. Mesoscale eddies have also been related to the dynamics of large-scale motions in NORPAX.

OCEANIC COUPLING WITH THE ATMOSPHERE

Many of the IDOE results in physical oceanography are related to air–sea interaction, climate, and energy exchanges. NORPAX has contributed significantly to the establishment of the existence of large-scale sea-surface temperature anomalies in the North Pacific. Correlations between the anomaly patterns and climatic variations over North America have been demonstrated. These results have stimulated further studies of anomaly formation and the role of these anomalies in global atmospheric fluctuations. In its equatorial-region studies, NORPAX has focused on the Niño phenomenon, which appears to be one symptom of the basinwide response to both the atmosphere and the ocean. A high correlation is found between the southern oscillation and the Niño phenomenon.

The large-scale atmospheric forcing of local upwelling observed during

CUEA has raised important questions about the relative roles of remote and local events. Connections have been found between variations in the ecosystem and circulation fluctuations, as noted in the discussion under biological oceanography. ISOS results have demonstrated a relationship between surface winds and currents in the Drake Passage. In addition, estimates have been made of meridional transports of heat. These studies are providing a better understanding of the role of the Southern Ocean in global climate. In POLYMODE there is evidence that eddy-driving may be an important mechanism for surface heat fluxes.

Studies of ocean currents have been closely related to those of physical oceanographic processes during the IDOE. The extensive results of ISOS relating to the Antarctic Circumpolar Current have altered our basic concept of its structure and variability. Long-term arrays of current measurements have defined patterns of spatial correlation in the Drake Passage. MODE/POLYMODE results suggest that midocean eddies interact with the Gulf Stream and the North Equatorial Current. CUEA has provided new understanding of the flow regimes of surface jets and undercurrents related to upwelling areas. For example, surface coastal jets and poleward undercurrents are found in all coastal upwelling regions.

THEORETICAL STUDIES

There has been a component of theoretical studies in all the physical oceanography programs in IDOE. CUEA has developed models of both physical and biological processes involved in upwelling. One interesting result is the common occurrence of upwelling patches near the heads of underwater canyons and equatorward of capes.

NORPAX has carried out theoretical studies on the role of Rossby waves in the interior of the Pacific and Kelvin waves in the equatorial region. Many of the longer-period fluctuations of the main thermocline in the subtropical Pacific can be explained as internal Rossby waves.

The MODE/POLYMODE projects have developed extensive quantitative models. These may be categorized in two ways: as process models, in which the processes of horizontal and vertical convergence and propagation of energy are studied, and as eddy-resolving general circulation models, in which the interaction of eddies with the mean field of circulation is examined.

ADVANCES IN TECHNIQUE

Finally, all the physical oceanographic programs in the IDOE have made significant advances in techniques used to study the ocean. The CUEA studies have integrated the use of research vessels, aircraft, moorings, coastal stations, and satellites.

MODE and POLYMODE have evolved synoptic mapping techniques in space and time. Our understanding of the application of objective analyses to oceanography has been improved, and reliable instrumentation for long-term, low-frequency measurements in the deep sea has been developed.

The ISOS has had remarkable success in carrying out reliable long-term measurements in the hostile environment of the Drake Passage. NORPAX has developed methods for large-scale monitoring of the heat content of the ocean, in particular through the use of satellites and remotely tracked drifters.

BIOLOGICAL OCEANOGRAPHY

At the beginning of the IDOE program, it was apparent that the traditional individual-investigator approach was inadequate for the major oceanographic questions that required multi-institutional participation, relatively large sums of money over long periods of time, and unusually costly facilities. This was particularly true in the field of biological oceanography, where, traditionally, single investigators undertook narrowly defined, specific problems. A few years after the IDOE began, a Living Resources program was initiated. Its principal activity has been the Coastal Upwelling Ecosystem Analysis (CUEA), recently joined by the much smaller Seagrass Ecosystem Study (SES). Other biological studies have been conducted within the Environmental Quality program, particularly in the Controlled Ecosystem Pollution Experiment (CEPEX) and in studies of the biological effects of pollutants and of pollutant transfer.

COASTAL UPWELLING

The CUEA project has been a comparative study of coastal upwelling in four regions, off Oregon, Baja California, northwest Africa, and Peru. In each case, the objectives were to describe the mesoscale distributions of variables defining the ecosystem, to measure the dynamics of both biological and abiotic processes that govern the major properties of the system, to document interactions of physical and biological processes, and to develop conceptual and simulation models of the system that account for and predict their differing biological character.

Each coastal upwelling ecosystem was found to have unique characteristics. Relatively small differences in meteorological forcing, bottom topography, stability, and circulation had large consequences for the ecosystem.

Two features common to all coastal upwelling regions and important to their ecology are a surface coastal jet and a poleward undercurrent. The latter appears to provide an essential feedback mechanism for nutrients and for phytoplankton and zooplankton.

Off Peru, relatively weak winds result in large dinoflagellate blooms, which, following strong winds, are quickly replaced by typical diatom assemblages within a few days. There appears to be an optimum mix of windy and calm conditions, with winds of sufficient strength to drive upwelling of nutrients yet not so strong as to create a deep mixed layer, and with calm periods in which phytoplankton are confined to a shallow surface layer with sufficient light for photosynthesis. The mix is more suitable for sustained high productivity off Peru than off northwest Africa. However, off Africa, the presence of a wide, shallow continental shelf and the associated cross-shelf circulation pattern lead to a more complex food web than exists off Peru.

Large-scale phenomena occurring far away in the central ocean play an important role in regulating productivity in the Peruvian coastal region. Such phenomena may, for example, cause the water of the poleward undercurrent to be warm and nutrient-depleted, so that even strong upwelling does not lead to high productivity. When the large-scale, remotely driven variations in circulation override local forcing, as in the 1972 and 1976 El Niño events, resulting changes in the biological character of the coastal ecosystems can have disastrous consequences for the anchoveta stocks.

CONTROLLED ECOSYSTEMS

In the ocean, the generation times of phytoplankton and zooplankton are on the order of days to months, while their distribution scales are kilometers to tens of kilometers. It is obviously impossible to conduct controlled experiments of such dimensions, and changes in pelagic populations can best be assessed experimentally in isolated parts of the environment. If the container is too small, however, the results of the experiments may bear little relation to the real ocean. Attempts to achieve adequate yet manageable size in experimental containers led to the CEPEX project, in which plastic containers 9.5 m in diameter and 29 m in length, with a calculated volume of 1700 m^3, were immersed in the waters of Saanich Inlet, British Columbia. Although controlled experimental systems of this size do not, of course, completely reproduce conditions in the ocean (for example, they eliminate horizontal advection and reduce vertical water mixing), they appear to offer a reasonable approximation of natural conditions, as evidenced by the food webs they reproduce.

A variety of experiments has been conducted in which the contained ecosystem has been stressed by the addition of pollutants such as copper, mercury, and petroleum hydrocarbons. Initial results show a similar sequence of events regardless of the pollutant added. With the introduction of copper, as the most vulnerable phytoplankton species collapse, tolerant bacterial populations develop rapidly. With oil, heterotrophic activity is initially depressed but recovers rapidly. Populations of centrate diatoms are replaced by some pennate diatoms and by microflagellates. Species succession occurs rapidly, so that most standing stock and rate measurements show little effect. Population structure and the succession of natural mixtures of plant species are more useful measures of pollution stress.

The sensitivity of zooplankton to metals is a function of their size, with smaller species and development stages being more vulnerable than larger animals. With sublethal doses of pollutants, the rates of egg production and of ingestion appear to be markedly reduced, although longer experiments will be required to discover how these responses affect the overall production and population structure of zooplankton.

Two other effects are worthy of note. Pollutants affect predators such as ctenophores as well as zooplankton prey, so that an indirect effect of pollution may be to reduce grazing pressure and thus to favor larger standing stocks of prey. Also, pollutants differ in the extent to which they are absorbed. The heavier metals and some components of oil probably have a greater impact on benthic than on pelagic organisms, since the particles on which they are absorbed sink quickly to the bottom.

In the pollution-stressed systems investigated, the sequence of events in phytoplankton and zooplankton succession is similar to that occurring over much longer periods of time in the surrounding natural waters in response to environmental changes such as light and availability of nutrients. Thus the controlled ecosystems, although stressed by "unnatural" agents, appear also to model the natural system in a useful way.

CHEMICAL OCEANOGRAPHY

An understanding of the nature and distribution of chemical species in the ocean requires studies of the composition of seawater, organisms, and sediment. Particular attention must be given to geographic, depth, and temporal variations of both inorganic and organic species. Studies of the relationship between these variations and various physical, chemical, and biological processes must be conducted. Equally important is the investigation of fluxes of chemical species both within the ocean water column and at the ocean boundaries. Compositional variations and fluxes must be thoroughly understood if they are to provide the input necessary for predictive models of the ocean and its interaction with the atmosphere, lithosphere, and biosphere.

The IDOE projects that have contributed directly to these goals include GEOSECS, CEPEX, CUEA, the Pollutant Baseline Project, the Pollutant Transfer Project, and the Galapagos Rift Project.

GLOBAL MEASUREMENTS OF CHEMICAL SPECIES

Measurements of the composition of the ocean, its organisms and sediments, and the marine atmosphere have been major elements of these and other projects. Through these efforts, much has been learned about the spatial and temporal variations of chemical species in the ocean. At the same time, confidence in the quality of the data has grown significantly.

Compilations of hydrographic and nutrient data are available from numerous past cruises undertaken by various institutions and countries. In general, because of their uneven quality, the nutrient and oxygen data of these compilations are of limited value except for broad regional comparisons. Prior to GEOSECS, no reasonably comprehensive global model of chemical properties of the ocean could even be contemplated.

This situation has been changed dramatically. It is interesting to note that the analytical methods employed for most of the previous nutrient determinations did not differ significantly from those used during the GEOSECS cruises. The difference is that the latter analyses were managed by a single group of scientists who paid strict attention to quality control; they were not individual efforts by many institutions characterized by little coordination, as before.

For the first time, as a consequence of IDOE projects, information has become available on the global ocean distribution throughout the water column of properties of the carbonate system; stable isotopes of carbon and oxygen, hydrogen and helium; fallout radioactive isotopes including ^{3}H, ^{137}Cs, ^{90}Sr, and several transuranic isotopes; and natural radioisotopes including ^{14}C, ^{226}Rn, and, to lesser extents, ^{228}Ra, ^{210}Pb, ^{210}Po, and ^{32}Si. The distribution and composition of particulate matter are also better understood.

These data, which could not have been produced without a closely integrated program between shipboard and shore-based operations, will form a basis for the development and testing of ocean models for years to come.

The simultaneous determination of many species, with differing chemical properties and boundary conditions, yields a data set that provides important constraints for these models. Examples of some of the important specific results already available from the GEOSECS observations include the following:

1. Documentation of the distribution of species of the carbon dioxide system in the ocean. This includes evaluation of several competing sets of apparent dissociation constants.
2. Demonstration, from tritium and carbon-14 distributions, of the extremely rapid renewal rate of deep water in the Atlantic Ocean, where most of the world's deep water is currently formed. Tritium and other fallout products, in association with the deep overflow water in the Denmark Strait, at 63° N, have already penetrated to 38° N in the western Atlantic. Following transient tracer distributions in the North Atlantic over the next decade or so should produce important information on the rate of renewal of the world's deep water, the relationship of the renewal rate to climate conditions, and the transport rates of chemicals in open-ocean conditions.

Although GEOSECS has been the principal effort in chemical oceanography, several other programs have resulted in major contributions. The CUEA project, whose principal objectives have been to understand the relationships between the physics of upwelling and ocean productivity, has shown how nutrient distribution in upwelling areas responds to changes in wind stress as well as to the biological processes operating there. The high biological productivity of the upwelling regions of the ocean and the consequent increase in the extent of oxidation-reduction reactions have effects that are evident far from the upwelling areas themselves. The CUEA studies contribute significantly to the understanding of the development of oxygen minimum zones in the ocean's interior.

CHEMICAL FLUXES AT THE SEAFLOOR

Preliminary data available from the Galapagos Rift Project have already deepened knowledge about the processes that brought seawater to its present composition and the origin of hydrothermal ore deposits in the deep ocean. The Manganese Nodule Project has collected and collated useful information on the composition and distribution of these nodules. Recent changes in the direction of the project promise to provide substantial data on processes involved in the origin of the nodules.

POLLUTANT INTERACTIONS AND TRANSFER

The interaction of chemical species with the marine biosphere has been investigated in several pollution-related projects. The CEPEX has provided basic information on the response of lower trophic-level organisms to stress by pollutants, including several trace metals and petroleum hydrocarbons. The short-lived project on Pollutant Baselines summarized the knowledge available in 1972 on the concentrations in seawater, organisms, and atmosphere of trace metals, halogenated hydrocarbons, and petroleum hydrocarbons. The pollutant transfer project has yielded much information on the composition of the marine atmosphere and has highlighted the importance of atmospheric transport processes in understanding the flux of materials to the ocean surface.

GEOLOGY AND GEOPHYSICS

Much of the geological work of the IDOE has been directed toward a fuller understanding of the geological processes at spreading centers, at continental margins, and in deep-ocean basins. Particular attention has been accorded processes of mineral concentration, since potential ore deposits are being created at spreading centers by the processes that generate new ocean crust. Continental margins are also sites of mineral deposits.

SPREADING CENTERS

Research submersibles were used in studies of the Mid-Atlantic Ridge and the Galapagos Spreading Center to permit a close look at details of these features. At the Mid-Atlantic Ridge, the primary focus was upon those processes that create new crust, while in the Galapagos it was on hydrothermal mineralization. Detailed studies in the former region showed that midridge volcanism is largely confined to a zone no wider than 100–200 m. Volcanic activity seems to be centered on a series of elongated hills down the center of the inner rift valley and is episodic in occurrence, with a period of about 10,000 years.

In rocks from the midridge rift, gradients of major element concentrations across the rift and variability of chemical composition are nearly as large as they are anywhere in the oceans. Several types of rocks—some of them primitive basalts formed at high temperatures, others formed at lower temperatures—appear to be differentiated from the primitive types. The differentiated rocks exhibit a symmetrical and gradual compositional change across the rift.

The extent of manganese coating suggests that the youngest rocks have formed in the midrift hills; older rocks are found near the margins of the rift valley. These relations may indicate that the diverse lava types erupted from a shallow-zoned magma chamber through fissures distributed over the width of the inner rift valley and elongated parallel to it. Differentiation occurred toward the margins of the magma chamber.

Associated investigations of fracture zones in the vicinity revealed that, although the zone, as indicated by sheared rock, was about 20 km wide, the currently active portions are less than 1 km wide. The fracture zones are areas of high microearthquake activity and relatively high heat flow.

Preliminary investigations of the Galapagos Rift using deep tow gear indicated areas of hydrothermal activity. Subsequent dives by the deep submersible *Alvin* were made to four major hydrothermal areas and to three apparently extinct hydrothermal vents. Equipped with temperature and chemical sensing devices, scientists aboard *Alvin* discovered that water emanating from the bottom was as much as 15°C warmer than the surrounding bottom water. High concentrations of hydrogen sulfide characterized the hydrothermal solutions. Sediment mounds, apparently formed by hydrothermal activity, surrounded some of the vents. Heat-flow measurements in these sediments indicated pore water temperatures up to 12°C in the upper portions of the mound. The most surprising discoveries made at Galapagos vents were biological. Unique communities of living organisms, including clams, mussels, limpets, crabs, pogonophora, and worms, were discovered living in the hot water from the vents. Sulfide-oxidizing bacteria appear to form the basis of the food chain for these communities. The episodic nature of some of the hydrothermal emanations was suggested by large accumulations of dead organisms where heated water no longer issues from the seafloor.

CONTINENTAL MARGINS

Most mineral production from the ocean today occurs beneath the continental margins. Virtually all the oil, gas, and sulfur extracted from the oceans comes from these regions. Continental margins are classified as active or passive; active margins occur at the edges of converging plates; passive margins, at some position in the interior of plates. Passive margins mark the original site of plate spreading and have assumed an intraplate position by the addition of new crust at the spreading centers. IDOE projects have studied passive margins on both the east and west sides of the Atlantic; active margins of Chile and Peru have been investigated as part of the Nazca Plate Project. More recently, studies in the Southeast Asia Tectonics and Resources (SEATAR) project have included active margins of Southeast Asia as their area of concentration.

Studies of Atlantic continental margins were designed to discover how and when Africa separated from South America, to determine the subsequent history and development of continental margins and the adjacent deep seafloor, and to locate possible economic mineral deposits. The African margin includes a number of large sedimentary basins containing more than 4 km of sediments, with many structural features suitable for oil and gas traps. Three belts of diapirs were discovered beneath the continental shelf and continental slope off northern Africa and in the region of the Niger Delta and off the mouth of the Congo River.

Gravity and magnetic studies along the margins of southern Africa and South America have improved the prespreading fit of the two continents, eliminating most gaps and overlaps. Diapiric structures similar to those off West Africa were encountered off the southern Brazilian shelf. On both margins these structures lie seaward of coastal basins, which are known to contain salt deposits. By fitting together the seaward edges of the diapir belts, a reasonable picture of the early stage of Atlantic Ocean opening is produced. This opening occurred about 150 million years ago.

As part of the Nazca Plate and SEATAR projects, studies have been made of active margins that integrate plate processes from crustal genesis at spreading centers to crustal destruction in subduction zones. Nazca Plate studies have focused on mineralization at spreading centers and transport of minerals to subduction zones. Metalliferous sediments of the East Pacific rise are derived both from hydrothermal activity and from seawater sources.

Chemical studies of basement rocks show similarities of composition within crustal segments and compositional differences between segments, differences that may ultimately affect the type of mineralization occurring above subduction zones in the mineral deposits of the South American continent. The Nazca Plate Project was the first to demonstrate accretion of oceanic plate and trench deposits to the continental block of South America. Segmentation of the Nazca Plate occurs in and landward of the subduction zone and must have existed long enough to be reflected in the geology of the Andean region and in current seismicity and volcanism.

MANGANESE NODULES

Manganese nodules of oceanic plates are the subject of the Manganese Nodule Study Project. Initially, the project summarized existing data and then embarked on field studies in a portion of the northern equatorial Pacific characterized by extensive deposits of copper- and nickel-rich nodules. Bottom-mounted equipment was used to monitor sedimentation and benthic organisms in areas of manganese nodule concentrations.

Microscopic structures on nodule surfaces and in their interiors may have been made by protozoans. This discovery re-emphasizes the importance of understanding the role of biological agents in the growth of deep-sea nodules. Postdepositional recrystallization destroys evidence of these organisms and results in minerals rich in copper and nickel. Investigations of sediment pore water of biogenic oozes show that manganese and copper are enriched 10 to 25 times in pore waters over their concentrations in bottom water. Pore waters of other sediment types, such as red clay, were found to be relatively low in manganese and copper. Causes for these variations are inadequately known at this time.

PALEO-OCEANOGRAPHY—PALEOCLIMATE

Marine geologists have also participated in interdisciplinary studies of paleo-oceanography. Climatic changes that have occurred in the past million years are the topic of investigations under the Climate Long Range Investigation Mapping and Prediction Study (CLIMAP). This study seeks better understanding of the physical mechanisms that cause major variations of climatic scale in the atmosphere and ocean. Deep-sea sediments were analyzed in order to reveal the geological record of the ice ages. Sediments that include the remains of organisms that lived in surface waters are particularly valuable. These biogenic sediments accumulate at a relatively constant and continuous rate, uninterrupted for hundreds of thousands of years over the entire ocean. From the remains of these organisms, inferences can be made concerning the temperature and circulation patterns of the surface waters, the chemical nature of the bottom water, and, to some extent, the distribution of sea ice.

Initial CLIMAP investigations have emphasized oceanic circulation and its atmospheric counterpart during the time of the last major glaciation, 18,000 years ago. An ocean circulation model has served as a base for derivation of a global atmospheric model. These models have made it possible to infer the ice-age climate and to compare it with that of today. The mean air temperature during the last glaciation appears to have been only 5 degrees cooler than at present. This estimate of the magnitude of the largest climatic change ever to occur during the last million years provides a basis for judging the impact of any changes in future climate that might occur naturally or as a result of man's activities.

A theory, for which detailed calculations were published by the Serbian Milvtin Milankovich in 1930, assumes that changes in the geometry of the earth's solar orbit result in changes of the seasonal and latitudinal distribution of solar energy and hence in climatic variation. Milankovich postulated that the three periodic variations in the orbit occurred as a result of changes in the position of the planets in the solar system as well as in variations in the eccentricity of the earth's orbit. These variations occur at average periods of about 100,000, 41,000, and 22,000 years, depending on the factors involved. If the Milankovich theory is correct, each of these periodicities should be reflected in the climatic record. On the basis of two deep-sea cores from the southern Indian Ocean, CLIMAP researchers have discovered climatic periodicity so close to that predicted that the Milankovich theory appears to be confirmed. CLIMAP studies have also described the movement of the North Atlantic Polar Front during the last major climatic cycle and have shown that at times the rate of movement averaged more than 1 km per year over several thousand-year periods. This suggests that a major change in climate from a full glacial state to an average interglacial state can occur in less than 3000 years.

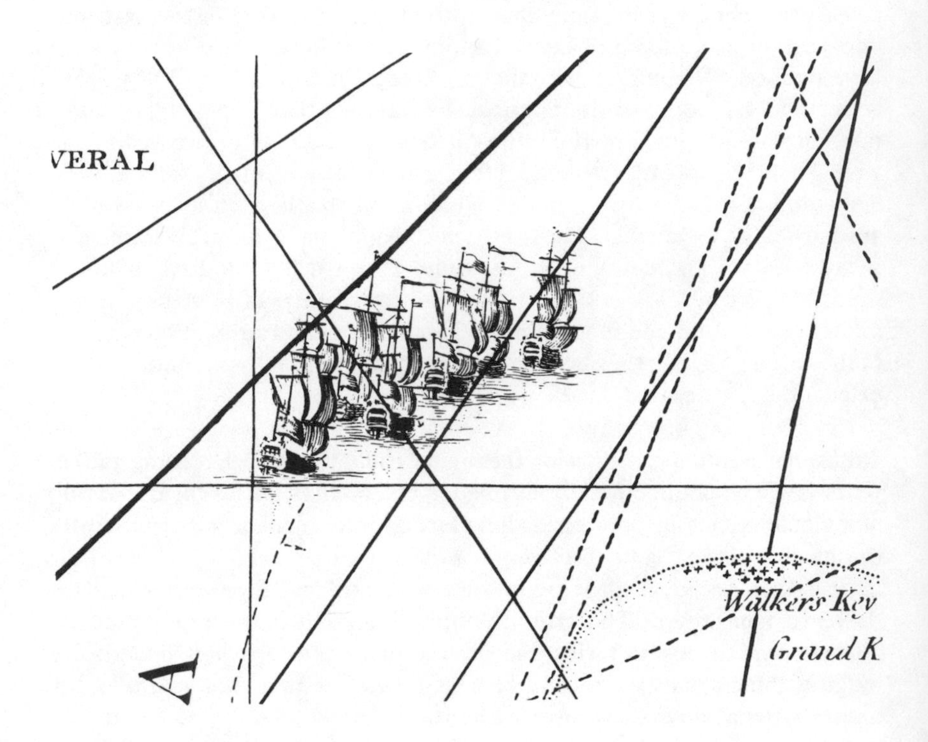

3 Marine Research and Society

FUNDAMENTAL AND PROBLEM-ORIENTED RESEARCH

Although academic scientific research is usually directed toward gaining knowledge and understanding, the results of such research are usually applicable, either in the near or long term, to the solution of some societal problem.

Whether research can be considered "pure" or "applied" depends in large part on the motivation of the investigation. For the purpose of our discussion, it is useful to distinguish between the extremes—*fundamental* research, the primary objective of which is to solve an identified scientific problem, and *application-oriented* research, which is directed toward the solution of a specific operational problem.

In developing research plans, such as the sample projects described in Chapter 4, scientists are chiefly interested in learning how the ocean operates. This desire to satisfy curiosity about nature characterizes all fundamental research and will be reflected in proposals submitted to the post-IDOE program.

Most ocean research supported by the National Science Foundation, including that of the IDOE program, can be characterized as fundamental research. Yet when the IDOE was established, its objective of achieving more comprehensive knowledge of ocean characteristics and more profound understanding of oceanic processes was linked to the more effective utilization of the ocean and its resources. This linkage has affected the organization of the IDOE office into four divisions identified with different areas of application. In practice, the IDOE investigations have leaned toward fundamental research.

We consider that the link with eventual application has given a useful focus to the IDOE and thus have proposed (see Chapter 5) that the post-IDOE program should continue to emphasize such a relationship and that the emphasis should be broadened to include protection of the marine environment and the forecasting of weather and climate. Meeting these objectives through a program of fundamental research is proposed for a number of reasons, including the following:

1. The context and significance of applied problems continue to change with the increase in the variety and intensity of ocean use. Whatever the problems of the moment may be, their solution eventually will be based on fundamental knowledge.
2. Oceanic phenomena and processes are complex and interactive. Interpretation of the results of application-oriented research usually requires an understanding of the underlying processes that can only be obtained by fundamental research.
3. Fundamental scientific knowledge is essential for predicting the consequences of alternative decisions on ocean use.
4. Fundamental research supports that which is application-oriented by generating new ideas and methods that can be applied to more immediate problems.

Many mission-oriented agencies deal with applied research. But fundamental research of the sort discussed in this report depends primarily on funding by the NSF. The development of new and fundamental knowledge is a central activity of academic institutions, and the NSF has a principal responsibility to promote fundamental research. Thus we are proposing an NSF-funded academic program of fundamental research. Elements of this program will usually arise because scientists wish to solve scientific problems. As discussed in this chapter, that approach is necessary for the ultimate solution of important operational problems.

While more fundamental research supports application-oriented research, we are particularly concerned with the role of large-scale, cooperative research such as that sponsored by the IDOE. Two of the most important benefits of such research are these:

1. Projects of such scope make possible interaction among disciplines, institutions, agencies, and countries commensurate with the complex nature of the problems addressed.

2. Such projects foster the education of scientists prepared to cope with the increasingly complex problems of the future.

Continuing an IDOE-like program into the 1980's can be expected to generate new scientific knowledge of the ocean and its interactions with the land and the atmosphere and to contribute to a rational basis for understanding and governing man's activities as they relate to this marine environment. In view of the ever-increasing pressure of man's activities on the ocean environment and its resources, continuation and expansion of such a program are, in our view, justified.

MARINE-RELATED PROBLEMS OF SOCIETY

While the post-IDOE program should emphasize fundamental research, the areas of eventual application of research findings should be kept in mind. Some indication of their importance and variety is given below.

Seventy percent (2.7 billion people) of the world's population lives within 200 miles of the ocean. Many people are nourished by the food that the ocean yields and the recreation that it provides. An ever-increasing fraction of the oil and gas consumed by society is obtained offshore; most is transported by sea. The ocean is the ultimate receiver of human and industrial wastes. It strongly influences weather and climate and plays an important role in national security.

The sea provides a significant percentage of the animal protein (estimated to be as much as 19 percent in developing countries) ingested by the world's population, and, as population expands, so does the importance of this food source. During each of the last several years, total landings of marine species have been on the order of 60 million tons. Important quantities of some species, such as squid, are only slightly exploited, and enormous stocks of unconventional species, such as krill, may, if technological problems and ecological questions can be resolved, yield as much additional food as present landings of all other species.

In order to produce on a sustained basis the greatest amount of food or to maximize other predetermined benefits, fisheries must be managed with a knowledge of the exploited stocks and their responses to the fishery and to environmental changes. The interactions among species are poorly understood; present single-species management, therefore, is far from the desired holistic, ecosystem approach. One of the more intriguing scientific problems is to predict the natural fluctuations in recruitment to important resource populations. Such predictions require elucidation of the interactions between the atmosphere and the ocean, of the resulting density structure and fertilization of the surface layer, and of the ways whereby variations in the physical environment are propagated through the food web.

To date essentially all of the world's energy has been obtained from the burning of fossil fuels—coal, oil, and gas. Since World War II, the proportional contribution of oil and gas has increased enormously, while a new energy source, the nuclear reactor, has gradually become important. With the realization of the limitation of land sources of oil and gas, there has been a shift to reservoirs on the outer continental shelf (in 1975, 16.4 percent of the U.S. supply of crude oil came from such reservoirs). Nuclear operations and oil and coal production, transport, and utilization pose some danger, as yet imperfectly understood, of environmental degradation; land-based energy supplies are limited. Thus the search has been initiated for other usable concentrations of solar energy, such as ocean waves and ocean-temperature differences.

Most minerals are obtained from land, although there is a growing supply of sand and gravel from the continental shelf where extensive placer and phosphorite deposits may eventually be exploited. The mining of ferromanganese nodules from the deep seabed is now under development. These nodules will likely become an important alternative source of nickel, cobalt, and copper. Other potentially valuable mineral deposits have been found in rift zones. The metalliferous brines in the Red Sea are an example.

The search for offshore petroleum and mineral deposits has employed geophysical methods and has been based on geological and geophysical knowledge arising from scientific investigations. During the past decade the knowledge base has been transformed by the new understanding of continental drift and the associated theories of plate tectonics, which arose from a series of large-scale academic investigations both within and outside the IDOE. New concepts concern the origin and development of marine basins, their accumulation of organic matter, and their stability during geological time. These have a bearing on the conditions for formation of petroleum hydrocarbons, have recently been linked to the generation of metalliferous ores, and may eventually be important in guiding disposal of nuclear waste.

Large-scale ocean research relates to exploitation as well as to exploration of nonliving resources. Wind, wave, and current profoundly affect conditions of petroleum extraction on the continental shelf; studies of air–sea interaction will assist in the prediction of these phenomena. How the offshore environment is understood and dealt with will continue to have a major impact on the financing, operation, and safety of the offshore petroleum and mining industries. In turn, the prediction and control of the impact on the marine environment of these operations, for example the consequences of major oil spills, also depend on the findings of large-scale investigations, both on ocean circulation and on pollution effects.

All nations depend on others to some degree for their supply of food, fuel, raw materials, and manufactured products. The great bulk of these materials is carried by ships. Although marine transportation has not been closely associated with oceanographic research, it is clear that both the safety and the economy of shipping are affected by weather, sea state, currents, and ice; forecasting of these conditions depends on scientific knowledge of the ocean. Understanding and controlling the effect of shipping on the quality of the marine environment also have an important scientific component.

Military security has always been provided in large part by sea power. Protection of vital shipping links is an ancient and continuing aspect of this security. Only within the last few decades has the ocean been used for concealment of the nuclear deterrent. A significant focus for oceanic research, therefore, has become the acoustic properties of the ocean as they affect the detection of submarines. The elucidation of these properties requires a thorough understanding of physical processes in the sea. This research has important applications in other fields, for example, in the detection, tracking, and assessment of fish schools and in the exploration for seabed minerals and petroleum deposits.

Throughout history, ports and other facilities and structures have been constructed along the coast. Major consumers of oceanographic information have come to include engineers and firms engaged in major structural design, power companies using the thermal capacity of coastal waters, and designers and builders of offshore power plants and tanker terminals, offshore platforms, and breakwaters. Much of the required information can be obtained by small-scale, application-oriented research. But the post-IDOE program could make an important contribution from improved concepts and numerical modeling of coastal processes and the development of better instruments. Such studies should lead to the installation and operation of monitoring systems, in which remote sensing will certainly play an important role.

Protecting the marine environment from pollution from ships and from the land, via rivers, runoff, or the atmosphere may require the control or reduction of these inputs. The cost of this protection can be so high that an industry is no longer economical or its productivity is seriously reduced. We need to know the fates and effects of pollutants in the ocean in order to evaluate their importance and to determine tolerable levels of their concentration. Without this knowledge there could be irreversible damage to the marine environment and its living resources or, conversely, industry and agriculture might be required to operate under unnecessarily stringent and uneconomical conditions.

Although much of the necessary pollution research must be done in the laboratory, the interaction of processes can best be studied in the sea. The

CEPEX project of the IDOE, in which plankton ecosystems are isolated in huge containers suspended in the sea and inoculated with pollutants, exemplifies a large-scale, multi-institutional approach likely to produce fundamental scientific information as well as essential insight into pollutant effects.

Large-scale studies of ocean circulation and mixing, which determine pollutant transport and dispersion, are also vital to the ultimate protection of the marine environment. For example, the safe operation of coastal and offshore nuclear or conventional power stations, or of offshore petroleum installations, requires detailed information on and an understanding of transport and mixing processes in the coastal zone and their exchanges with the open ocean, particularly under storm conditions. Large-scale research on the oceanography of the continental shelf and slope is needed to determine the environmental impact of these activities. Such research will tie together and permit interpretation and extrapolation of the short-term efforts now prevalent in this zone.

An even broader problem is posed by proposals to store long-lived radioactive or other toxic wastes on or in the floor of the deep ocean. Here the requirements are geological and geophysical information and a detailed understanding of ocean mixing over global distances on long time scales.

Finally, the ocean plays a leading role in the development of many types of forecasts—of weather, of the onslaught of hurricanes and storm surges, of the abundance and availability of fish, of the consequences of man's activities on marine ecosystems. The critical role of the ocean in affecting climate serves as an example.

The climatic state has been subject to natural variations since the evolution of the ocean and the atmosphere, and these variations have affected life since its creation. Climatic changes over the millennia have forced man to adjust his activities and to move, when necessary, to more favorable environments. However, as he became more numerous and began to crowd the livable regions of the planet, his options to relocate were sharply diminished. Adjustment to unfavorable conditions tends to expend ever larger amounts of limited resources of energy and water. This technological approach has always strained and now usually exceeds the resources of the poorer countries, where most of the world's population lives.

In recent years, the importance of climatic change has been underlined by events such as the prolonged drought in the Sahel region of Africa, the recent coincidence of severe winters in the United States with the growing shortage and increasing cost of fuel, and major crop failures in many parts of the globe. Further stimulus to an attack on the problem has been the increasing recognition that climate may be affected by human activities. A case in point is the awareness of the potential effects of carbon dioxide released by the burning of fossil fuels. Climate forecasting will not alleviate the problems of adjustment to unfavorable conditions, yet it will permit orderly planning of the massive modifications of human activities that might eventually become necessary.

Although it has not been ruled out that nonperiodic variations in solar radiation may cause climatic changes, it is generally accepted that much of the variance is contained within the atmosphere and the ocean. The ocean is particularly important since it is the principal reservoir of heat and moisture that are exchanged with the atmosphere both locally and on a long-term and global scale. Furthermore, the ocean is a major sink for carbon dioxide; its capacity will determine the consequences of increased carbon dioxide production.

The sedimentary remains of marine organisms record past oceanic conditions extending well beyond the limits of human history. Such records enlarge our perspective of the range and nature of natural events and can contribute to future projections of climatic behavior in the face of both natural and human forces.

In the light of these factors, long-term and large-scale cooperative ocean research can be expected to play a central role in the new and expanded program of climate research now being planned.

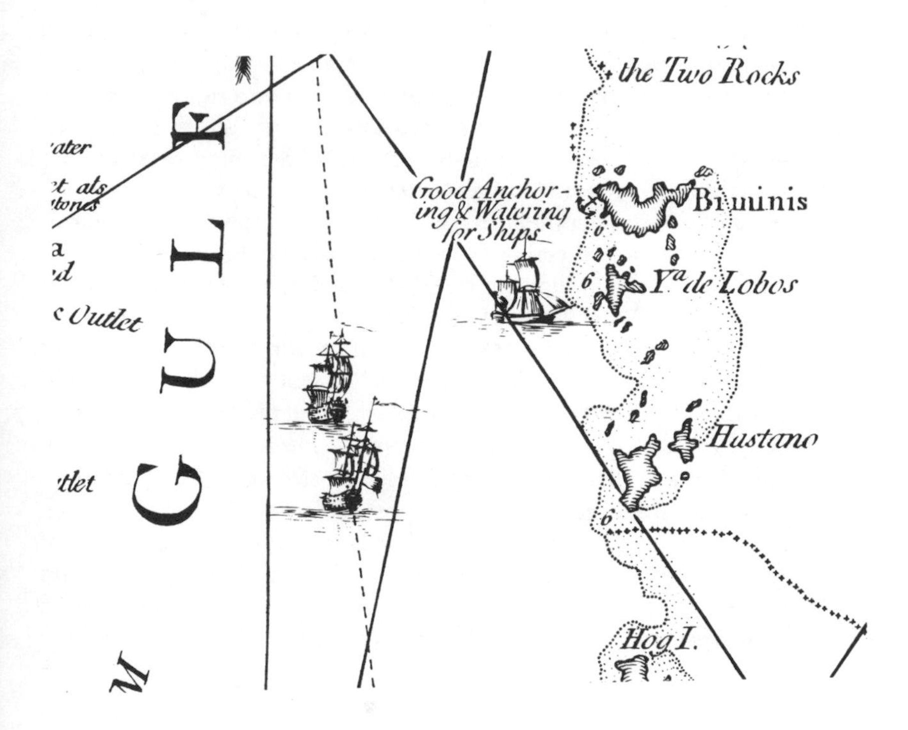

4 A Science of Complex Interactions: Oceanographic Opportunities for the 1980's

INTRODUCTION

Our consideration of the relation between marine research and society led to the conclusion that a program of fundamental research is necessary to generate the new scientific knowledge needed to provide a rational basis for understanding and governing man's activities as they affect his marine environment. We now examine the nature of the fundamental research that seems likely to be most fruitful during the coming decade.

Until recently, it was common to consider oceanography as the application of various scientific disciplines to the study of a common environment rather than as a science in its own right. It is now apparent that much of marine science lies at the interface between the conventional disciplines. As oceanography has moved from exploration and description to experiment and the testing of hypotheses, it has become evident that complexity and interaction are at the heart of oceanic phenomena and processes. It may well be that, as a science of complex interactions, modern oceanography provides a useful model for society in its study of other complex systems.

In this chapter we discuss research opportunities for the 1980's as seen by scientists in the several oceanographic disciplines and give examples of the interdisciplinary framework within which many of these opportunities might be fitted. Interdisciplinary projects may arise, for example, because of geographic proximity or because integrated studies of common oceanic processes are desirable. Some examples of specific cooperative research projects are then presented.

Through the IDOE program of the present decade, many oceanographers gained experience in mounting large-scale cooperative efforts to study ocean problems. Most of these activities were successful. The resulting data sets and knowledge about the ocean could not have been obtained in any other way. We now have a more realistic understanding of oceanic processes and their interactions and have developed models to describe them.

It seems clear that disciplinary studies will continue to play a major role. For example, within physical oceanography the knowledge gained from the study of eddies is fundamental to understanding the phenomena connected with ocean circulation. Cooperative efforts on this particular subject will be required through the 1980's and perhaps longer so that the results can be applied generally.

Reports of the disciplinary workshops give evidence that we may be ready now to attempt a few really *inter*disciplinary studies, where knowledge from one discipline enhances the study of another. Such research will require, by its nature, simultaneous measurements, communication among workers, and feedback, both positive and negative. In contrast, *multi*disciplinary research combines the efforts of several groups of workers and approaches that, while compatible, are not necessarily interdependent for conceptual understanding or success of the project.

Truly interdisciplinary studies are a relatively new venture for oceanographers. During the IDOE, biologists, chemists, and physical oceanographers worked together in the CUEA project, and geologists, physical oceanographers, and meteorologists cooperated in CLIMAP. Future studies will benefit from their experience. It is apparent that interdisciplinary studies often require additional scientific effort, beyond the extra management required, if mutual understanding is to lead to scientifically fruitful experiments.

DISCIPLINARY OPPORTUNITIES FOR THE 1980's

There are manifold research opportunities that might be exploited during the 1980's. As research proceeds, old questions acquire additional importance while new questions continually arise. Some that emerged during the workshop discussions are enumerated here. Although they are presented by discipline, it is clear that some fit well within interdisciplinary projects, as will be discussed in Chapter 5.

PHYSICAL OCEANOGRAPHY

Estuarine–Shelf Dynamics

What are the specific physical processes that transport saline shelf water into and freshwater out of an estuary? How can the vertical and horizontal mixing processes involved in the salt (and mass) transport be described, quantified, and modeled? How do these processes affect sediment transport?

Continental Shelf Dynamics and Shelf–Ocean Coupling

What dynamics govern the wind-driven and lower-frequency transient motions over continental shelves? Here vertical momentum transfer in the surface and bottom boundary layers must be described and parameterized. How does low-frequency variability influence shelf circulation?

Western Boundary Region Dynamics

What role do energetic western boundary currents play in (1) generation of midocean mesoscale variability, (2) interior recirculation, and (3) transport of heat?

Midocean (Interior) Dynamics

Can the steady horizontal (and vertical) circulation within a midocean subtropic gyre be computed from a knowledge of the density field?

Large-Scale Atmosphere–Ocean Coupling

How does the ocean interact with large-scale, low-frequency atmospheric forcing?

BIOLOGICAL OCEANOGRAPHY

Climate Variability and Productivity

How are temporal climatic changes reflected in variations in the kinds, quantity, and fate of phytoplankton, i.e., primary productivity of the seas? Phytoplankton is the basis of the marine food chain, and changes in other marine populations may therefore ultimately depend on the link between atmospheric climate and the ocean.

Physical Forcing of Species Succession

How do winds, storms, currents, upwelling, and other oceanic processes affect marine organisms so that one species is succeeded by another? How are organisms in the sea linked by trophic interactions? Understanding the effects of these interactions on the food web is essential for describing the marine ecosystem and its relationship to usable food resources.

Biological Interactions among Species

What are the effects of behavioral and other interactions among species, e.g., competition, predation, and chemotaxis, on the functioning of ecosystems in the sea?

Trophic-Level Coupling

What are the steps in the flow of energy through the food web from the initial absorption of sunlight to the eventual human consumption of fish flesh? How do the processes of energy flow differ in different oceanic habitats? Grazers, predators, and decomposers are all elements of this system.

Community Structure

What are the evolutionary and short-term adaptations to the physical, chemical, and biological forces in the sea that determine community structure there?

Patchiness

How does temporal and spatial heterogeneity in the distribution of plants and animals affect the concentration, survival, and availability of marine organisms? How do such concentrations affect the predators, associates, and competitors of an organism?

Recruitment

What are the key factors and timing in the development of marine organisms as they pass into the adult phase? How do environmental variations affect the abundance of these recruits?

CHEMICAL OCEANOGRAPHY

Water-Column Fluxes and Reactions

To what extent and on what time scales do the introduction, formation, and dissolution of particles control the composition of the ocean? How do the formation, composition, and vertical flux of particles vary with surface productivity, biomass structure, geographical location, and major ocean-circulation features? What are the reactions to the particles during their passage through the water column?

Seafloor Fluxes and Reactions

What are the compositional, geographical, and time variations in particle fluxes to the seafloor? Do processes occur at the seafloor that absorb or precipitate chemicals from solution? What is the nature of inorganic and organic reactions, what are the fluxes across the seafloor boundary, and how do they relate to or depend on biological or physical factors and sediment type?

Fluxes from the Continent to the Ocean

What factors determine fluxes of dissolved and suspended materials in major rivers? What are the sediment sinks and sources in estuaries? What are the sediment sinks and sources and the net fluxes across the continental shelf? How do biological processes affect the transport? What are the nature and magnitude of aeolian fluxes, and how are they distributed to the ocean?

Transient-Tracer Studies

What are the current rates of formation of deep water, and how rapidly do CO_2 and soluble contaminants penetrate into the ocean as indicated by transient-tracer studies?

Gas-Exchange Studies

How do exchanges of CO_2, N_2O, radon, and other gases vary with time, latitude, and sea-surface characteristics?

Tracer-Injection Studies

Can information on diffusion across the main thermocline be obtained through tracer-injection studies? What are suitable tracers for such experiments?

MARINE GEOLOGY AND GEOPHYSICS

Characteristics and Driving Mechanisms of the Deep Lithosphere and Asthenosphere

How can current models of the evolution of the lithosphere and asthenosphere and of the mechanisms that drive plate tectonics be validated? Can such studies define the evolution of plate parameters and constrain the possible rheology and driving mechanisms in the upper mantle throughout the life of existing oceanic plates (about 160 million years)?

Evolution and Variability of the Ocean Crust and Upper Mantle

What are the nature and variability of the oceanic crust and upper mantle? How are they formed, and which processes are important in their evolution? Which factors govern the emplacement and distribution of rocks at ridge crests? What controls compositional and structural variability at ridge crests? What produces mineralogic and structural changes with increasing age?

Structure and Evolution of Passive Continental Margins

How do passive margins—the boundaries between continental and oceanic crust within lithospheric plates—evolve, and what are the tectonic mechanisms responsible for this evolution? Can we determine and interpret the shallow and deep structure of passive margins during their prerifting, early opening, and mature stages?

Structure and Evolution of Convergent-Plate Margins

What are the processes involved in the formation and subsequent evolution of convergent margin systems (subduction zones, island arcs, and back-arc basins)? What are the rates and duration of convergence, and how do variations in lithospheric structure affect seismicity, volcanic activity, and structural evolution of convergent-plate margins?

Diagenesis at Depth

What biological, physical, and chemical processes modify sediments after deposition (normally referred to as diagenesis), and how are these related to processes in the benthic boundary layer? What can such studies reveal about the properties of consolidated sedimentary rocks and the chemical history of the oceans?

The Ocean's Role in Climatic Change over the Past 150,000 Years

What are the causes of the major climatic changes that appear in the geological record of the Quaternary ice ages? How can comparison of paleoclimatic observations with the quantitative predictions of physical theory contribute to the solution of this problem?

Climate over the Past 5 Million Years

What are the details of the oceanic history of the past 5 million years? The development and deployment of a coring technique to obtain detailed, undisturbed records of the top 100 m of the sediment column will greatly improve access to this information.

Changing States of the Ocean

What phenomena altered the circulation and thermohaline properties in the ocean over the past 150 million years? What has been their impact on climatic change and on the evolution of modern life?

AN INTERDISCIPLINARY FRAMEWORK FOR COOPERATIVE OCEAN RESEARCH

In recent years, we have become conscious of environmental problems with increasingly large space and time scales. Climate studies involve changes in circulation at entire ocean scales. Deep-sea disposal of nuclear wastes will concern very long time scales in chemical and geological processes. Increased national zones of management responsibility require that regulatory schemes for pollutants and fisheries be more comprehensive and deal with larger areas of the oceans. We must learn to measure and model intersecting aspects of ocean physics, chemistry, biology, and geology and geophysics as these interactions apply over long time and large space scales. This is a challenge for basic science in the 1980's and especially for interdisciplinary programs.

During the present decade, IDOE projects studied many areas but few (e.g., GEOSECS and CLIMAP) were devoted to the large time and space scales. Our outlook must now broaden. For example, the MODE project evaluated relatively rapid energy processes at scales up to 1000 km. CUEA studied nearshore processes, concentrating on the problems of basic productivity in upwelling areas with little direct work on the herbivorous fish. The CEPEX experiments on food chains are at the scale of 10 m. These small and mesoscale studies have been considered essential prerequisites for further work at large scales. Physical theories involve energy transfer between small and large scales and so require field studies at intermediate scales.

Much of recent biological oceanography has also been concerned with mesoscale plankton patchiness. Again, the thorough study of biological variations at these scales was needed before larger-scale environmental aspects of fishery management could be tackled. Similarly, problems of pollutant introductions concern the flux through a series of regions of increasing scale: horizontally from estuaries through coastal processes to the open ocean; vertically from a surface film through the upper mixed layer to deep water.

Thus, for societal as well as scientific reasons, interdisciplinary projects are necessary and must have a unified approach to the progression to larger scales in certain major geographical areas of study. There are numerous problems in attempting this approach—some conceptual, some technological. For example, the upper and lower boundaries of the ocean are of great interest for biological, chemical, and geophysical studies. We are overcoming the problems of working at the lower boundary, but, curiously, there are still major difficulties in measuring such processes as vertical shear and mixing in the upper 100 m, where knowledge of these dynamics is essential for biological studies. Again, the collection of long time-series data with fine (1–10 hours) resolution is now routine in physical oceanography but is, as yet, not possible in biology or chemistry, where such comparable data could greatly increase our understanding of the physical–biological–chemical interactions. The ability to make such measurements will be essential in the next ten years, and their development should be seen as a critical part of the attack on geographically diverse interdisciplinary problems.

The technical problems have created some difficulties in matching physical, chemical, and biological concepts. Thus there is a certain dichotomy between the study of temporal series in physics and spatial patterns in biology or in the distribution of trace constituents in the ocean. This can, in part, be overcome by theoretical studies that can extrapolate from existing data and can also indicate which joint physical, chemical, and biological measurements are required. Interdisciplinary modeling efforts can provide links within and between projects and so provide a framework for the apparent diversity of regional studies.

The longest time scales are encountered in the fields of marine geology and geophysics. Studies in these fields must interact not only with those in other marine disciplines but also with geological and geophysical investigations of the adjacent land and of the underlying solid earth.

The ideas that follow are only intended as examples of areas of interdisciplinary research opportunities. They, and the set of possible research projects that follows, serve to illustrate some of the important interactions and fruitful fields of future inquiry.

ESTUARINE AND COASTAL STUDIES

In a study of estuaries and their relation to coastal processes, an array of important scientific and societal problems can be attacked, utilizing all the major scientific disciplines that make up the field of oceanography. Physical oceanographers have proposed studies of estuarine and shelf dynamics, and of the role of estuaries and the continental shelf in oceanic mixing. Mesoscale processes, a few kilometers in size, are important in estuaries and the inshore zone.

Geologists have interests in sedimentary processes as well as in the structure and evolution of continental margins. A prime objective is to identify climatic indicators in rapidly deposited coastal sediments. Chemists are concerned with processes affecting fluxes of materials, and both geologists and chemists can contribute to an interdisciplinary study by providing environmental and sedimentological information to define important transport processes. In the coastal zone, chemical oceanographers are concerned with the accumulation of natural and man-made materials, including pollutants, rates of sedimentation, buildup of organic material, and fluxes of liquid and gaseous hydrocarbons from coastal marine seeps.

Biologists have identified five major unifying topics that must be studied in order to provide a basis for management of the coastal zone. These are (1) temporal and spatial changes in distributions of organisms, (2) species succession, (3) the role of each trophic level, (4) community structure and adaptation, and (5) the recruitment of key organisms.

Heretofore, studies of U.S. estuaries and coastal shelves have been local, uncoordinated, and focused primarily on immediate applied problems. Today there is widespread agreement among scientists, federal agencies, and local and state communities, all of whom agree on the need for a large-scale and coordinated fundamental study in this important regime.

Testable hypotheses now exist on the cause of large-scale effects of weather on the ocean, particularly on the coupling between oceanic and shelf circulation and the manner in which these events might affect fluctuations in basic productivity. The convergence of physical, biological, and geological hypotheses about teleconnections in the ocean reflects an agreement that studies of this size and scope may lead to a fuller understanding of the relationship between climate variability and biotic response. In the 1980's we must seek to elucidate the nature of the linkages involved in these processes.

Table 3 shows how the various oceanographic disciplines can contribute to multidisciplinary and interdisciplinary studies of estuarine, coastal, and continental shelf processes and their interactions with the open ocean.

EQUATORIAL DYNAMICS

A large portion of the variability of the tropical ocean is thought to be directly linked to the atmosphere and is amenable to remote monitoring, theoretical modeling, and even prediction. For example, the phenomenon of El Niño is a large-scale event in the atmosphere and ocean of the equatorial Pacific that strongly affects biological communities at the equator and along the eastern side of the ocean. It is heralded a few months before by a relaxation of easterly surface winds and increased rainfall in the western Pacific. Its effects then spread eastward and poleward along the eastern oceanic boundaries.

Physical oceanographers are intrigued by the wide variety of time-dependent circulations found only in the equatorial oceans. The propagation of events for great distances along the equator and to depths well below the surface is now predicted by theory. These horizontal circulation patterns can be directly measured in deep water with moored current meters and near the surface by satellite-tracked buoys; they are also suggested by remotely sensed sea-surface temperature. Vertical circulation patterns can be inferred from the distributions of nonconservative chemical species.

Biological oceanographers wish to describe and understand the dynamics of the rich equatorial populations and their relation to the changing environment. They are concerned with temporal and spatial changes in the distributions of organisms, species succession of phytoplankton as it affects survival of fish larvae and their ultimate recruitment as adults, and the relative role of each trophic level in the ecosystem.

The cycles of nutrient regeneration exhibit large fluctuations and may be related to particulate flux variations from the atmosphere as well as to the sources and sinks of nonconservative chemical constituents in the ocean. The equatorial thermocline is shallow, and the nutrient supply lies close to the surface. Small vertical excursions of the deeper waters can change nutrient concentrations at the surface enormously.

TABLE 3 Applications of Oceanographic Disciplines to Estuarine and Coastal Studies

Discipline	Estuarine–Coastal Processes	Shelf Processes	Shelf–Ocean Coupling
Physical oceanography	Salt intrusion (small-scale mixing, dispersion)	Freshwater extrusion (small-scale mixing, dispersion)	Subtidal shelf–ocean coupling (includes upwelling, warm-core rings, Calif. current, Equatorial Pac.)
	(Boundary-layer dynamics)	Boundary layer (surface and bottom dynamics)	Frontal mixing, stability, and dynamics
Chemical oceanography	→ Atmospheric Fluxes (Anthropogenic/Natural) →	→	→
	River and estuarine (particulate and dissolved material) fluxes	Reef dynamics → Hydroseep	→
	Chemical interactions and processes of sediment accumulation →	→	Climatic indicators
		← Shelf–ocean fluxes →	← Shelf–ocean fluxes →
Geological oceanography	Sediment sources and sinks, diagenesis and sediment transport → (Bottom boundary layer, turbidity currents, small-scale mixing by internal waves, canyons, mass sediment properties)	→	→
	←	← Paleoclimatic sedimentary and geomorphology processes (includes anoxic deposits and varved deposits)	
Geophysics	← Geological structures →	→	→
Biological oceanography	Larval contributions to recruitment →	Nutrient supply, primary and secondary production →	Pelagic and benthic populations

Equatorial ocean studies can profit greatly from interdisciplinary cooperation. The understanding of circulation patterns and of vertical and horizontal mixing and air–sea interaction in the tropical oceans is crucial to understanding both biological and chemical processes and their variability. Chemical and biological indices at the equator reveal circulation patterns that have not been directly measured. Understanding of the equatorial ocean system as a whole is a fascinating scientific problem as well as an important societal task.

THE SOUTHERN OCEAN

Problems of physics, chemistry, and biology are interlinked in the Southern Ocean. The biologist sees the region as an unusual habitat characterized by a short food chain and an abundance of nutrients, with an enormous potential for fisheries. Physical and chemical oceanographers recognize the possibility of relating this productivity to physical and chemical processes such as the formation of various water masses. By working together, physical, chemical, and biological scientists may be able to gain sufficient understanding of the total ecosystem to permit effective management of the living resources there.

The whole ecosystem is characterized by a variety of different scales, and each of these scales is related to different physical processes. For example, mesoscale plankton patchiness is probably related to water-mass boundaries and mesoscale mixing phenomena, whereas the general distribution of plankton is associated with the large-scale energy exchange and the behavior of large-scale currents.

Basic physical and chemical knowledge is required to study even the relatively simple communities of the Southern Ocean. Fundamental problems include the relation of the temporal and spatial distributions of phytoplankton to those of krill and other zooplankton, the roles of each trophic level in capturing and distributing energy, the importance and direction of species succession and the ways in which these may affect predators, and the factors leading to community structure and adaptation. Biological productivity and the distribution of krill in the Weddell Gyre almost certainly depend on the variability of the physical and chemical processes over the entire Atlantic sector of the Southern Ocean. For example, the large polynya (ice-free regions) in the Weddell Sea are of interest not only for bottomwater formation but also as regions for enhanced photosynthesis.

Processes of climatic scale also affect the biology. The oceans and atmosphere are coupled in the Southern Ocean on the long time scales (semiannual and longer periods) during which variations in the ice cover become

TABLE 4 Boundary-Layer Processes

Boundary	Chemistry	Biology	Geology	Physics
Air–sea boundary	1. Flux through the air–sea boundary (a) material (b) rates (c) processes 2. Production of particles 3. Gas exchange across boundary (a) CO_2 (b) N_2O	1. Fluxes through the boundary and energetics in the layer (euphotic zone) as they interact with (a) time and space changes in distribution of organisms (b) species succession (c) the role of each trophic level (d) community structure and adaptation (e) recruitment	1. Particle flux from terrestrial sources by (a) atmosphere (b) runoff from land 2. Particle production in boundary layer (euphotic zone)	1. Processes of energy exchange across boundary (a) heat (b) momentum (c) material 2. Surface-layer and water-mass formation
Bottom boundary	1. Flux through and reactions in boundary and media on both sides (a) deposition (b) resolution (c) diagenesis	1. Interaction of energy and material flux with the five biological processes, (a) to (e) above	1. Flux of energy and material: (a) deposition (b) erosion and redeposition (c) re-solution (d) upward flux of new material from earth's crust	1. Physical interaction water and bottom (a) friction (b) topography (c) energy conversion (d) electrical ionic transport

significant. The resulting air–sea–ice problem is crucial for studies of climate variability. For example, one of the prime questions is whether the growth of the ice pack is a positive or negative feedback in the polar heat balance. Its elucidation requires study of atmospheric forcing, large-scale ocean circulation, and ice-cover variability. Biological information on equivalent scales will be required to reveal the response of the ecosystem to this changing environment.

The composition of the world ocean, particularly as it concerns those chemical species that are strongly influenced by biological cycles, is significantly modified by processes operating in the Antarctic. Moreover, air–sea exchanges (particularly gas exchanges) have significant effects. For example, the Antarctic region provides venting for the bottom waters of the world ocean. It has long been known that bottom water is formed in the Antarctic; recent work has led to the hypothesis that intermediate water may be formed north of, as well as at, the Polar Front in selected regions of the Southern Ocean. Better understanding of these exchanges and water-mass formations will require research on surface mixed-layer dynamics, frontal dynamics, and deep mixing processes.

BOUNDARY-LAYER PROCESSES

In the surface boundary layer the ocean is forced by the atmosphere. A knowledge of processes in both atmospheric and oceanic boundary layers is essential to understanding fluxes across the surface. For example, how deeply does the ocean respond to storms? High wind speeds may account for most of the momentum transfer to the ocean, but we have few measurements taken under stormy conditions. A joint meteorology–oceanography study is required. Biologists also need to define mixing and transfer processes in the upper layer because of their effect on biological productivity. Chemists want to know about fluxes of climatically significant gases such as CO_2 between the atmosphere and surface ocean.

Chemistry, geology, physics, and biology must all come into play in studies of the bottom boundary layer. Chemists, biologists, and geologists wish to understand the nature, magnitudes, and mechanisms of sedimentation of solid phases. Geologists seek information on the physical, biological, and chemical processes that act to resuspend and transport sediment particles. Bioturbation is significant to mixing in the near-surface sediment and affects the rates of chemical reactions and fluxes of chemicals. Finally, the physics of the bottom boundary layer determines local mixing and transport and also constrains the large-scale circulation.

Table 4 shows the opportunities for cooperative studies at the air–sea and bottom boundaries in each of the several disciplines.

EXAMPLES OF POSSIBLE RESEARCH PROJECTS

The post-IDOE scientific program will arise from oceanographic opportunities such as those outlined above and will be elaborated in specific research projects. IDOE experience has shown the difficulty of anticipating in any detail the projects that eventually materialize. The sample projects given below illustrate some promising approaches. The set of examples is incomplete, and some are more fully developed than others.

In the preparation of these examples, consideration was given to their compatability with the criteria for post-IDOE projects discussed later (Chapter 5). Such investigations could be appropriately implemented in the coming decade because

1. The scientific community has become interested in the problem, has developed some experimental and theoretical understanding and capability, and is now ready to test hypotheses and models.
2. Solution of a relatively simple, yet fundamental, problem would open the door to tackling more complicated ones.
3. Consolidation of fragmented efforts in the study of important regions and processes is a recognized necessity.
4. Scientific information is urgently needed for the prediction, management, or control of human activities in areas of social and economic importance.

COASTAL AND ESTUARINE PROBLEMS

Fjord-Type Estuary Experiment

Problem Fjord-type estuaries are characterized by submerged sills that tend to block the flow of saline shelf water into the estuarine basins. Salt intrusion, deep-water renewal, and small-scale vertical mixing within the estuary are the dominant physical processes controlling the general circulation of these estuaries, which occur both in the United States and elsewhere. The proposed experiment is intended to measure and to provide a basis for modeling the processes that control the transport of saline shelf water into the estuary and vertical salt transport within the estuary.

Approach Instruments would be moored for long periods to measure (a) the mass and salt inflow over the sill, and the deep and surface flows at one or more sections across the estuarine basin, and (b) the transient shelf circulation just outside the estuary mouth and sill. Conventional hydrographic techniques and moored instruments would be used to measure the gross structure and variability of the temperature, salinity, and density fields near the sill and within the basin. This information, together with knowledge of the atmospheric forcing, river input, and boundary conditions at the sill, can be used with theoretical models to predict lower-frequency variations in the net horizontal and vertical salt transports within the estuary. The model predictions can then be tested against the observed horizontal circulation and vertical salt transports determined by the studies of small-scale mixing.

Comments Various groups of physical oceanographers would be involved in long-term measurements of current and density fields and in high-frequency observations of vertical salt transport. Recent experimental programs have developed instruments for the study of small-scale mixing, and the conceptual framework and physical modeling are topics of active theoretical work.

Bar-Built Estuary Experiment (Physical)

Problem Bar-built estuaries are shallow and have a horizontally complex topography with various channels and islands and with shelf and estuarine waters exchanging through openings (ports) in the bar. Tidal currents are strong, especially near the ports, and the estuary is well mixed vertically. Exchanges of water and salt between the estuary and the nearby region of the continental shelf vary over time scales of hours to weeks. They are driven by (a) rectification of tidal currents, (b) local winds, (c) freshwater inputs (especially when extreme runoffs are associated with tropical storms), and (d) low-frequency transient motions on the shelf driven by offshore eddy motions and/or coastal-trapped shelf waves. The experiment would determine those processes that govern net water and salt flux into the estuary, the time scales of their variations, and the manner in which small-scale mixing and advective processes adjust current and salinity (and density) fields within the estuary to boundary fluctuations in water and salt transport.

Approach Instruments would be moored to measure (a) water and salt flux through ports into the estuary, (b) tidal currents, and (c) mean subtidal circulation and its variability, both within the estuary and on the shelf outside the ports. Conventional hydrographic methods and moored instruments would be used to measure the spatial and temporal structure and the variability of the salinity (and density) field within the estuary. Off the East Coast of the United States, Gulf Stream eddies can be monitored to determine the large spatial-scale fluctuations; local measurements of shelf transients near the estuary would determine the local influence of Gulf Stream eddy variability on the flux into the estuary. A numerical model would be developed to predict the depth-averaged temperature, salinity, and current fields within the estuary, given the measured mass, momentum, and thermal energy input at the land, shelf, and surface boundaries.

Comment The physical experiment would involve the coupling of complicated theoretical and numerical models with a large, well-focused field program. It would also provide basic information for an associated biological investigation (see below). Studies of sediment transport in the estuary would also benefit from the determination of physical processes there.

Bar-Built Estuary Experiment (Biological)

Problem It is commonly believed, yet inadequately documented, that estuaries serve a vital function as nursery grounds for major fish and invertebrate populations of adjoining coastal waters. The estuarine habitat is subject to naturally fluctuating environmental conditions as well as to pollution, eutrophication, excessive sedimentation, and other influences of human activities. The problem is to determine the influence of estuaries on recruitment of coastal and pelagic species and the effects of human activities and natural forces on such recruitment.

Approach The investigation would depend heavily on the parallel physical experiment (see above), which would establish the character of fluxes and mixing within the estuary and between it and the adjoining coastal waters. For a few typical bar-built estuaries, integrated physical, chemical, and geological programs would determine the energy and material balance, the internal mixing regime, and the key recycling processes that regulate the quality of the estuary as a nursery. The biological component of the experiment would include, for appropriate species, studies of distributions and life histories, trophic-level interactions, evolutionary processes and community adaptations, species succession, and productivity. In short, the biology, food chain, and population dynamics of particular species would be determined and related to environmental variations and to exchanges between the estuary and the nearby open sea.

Comment The proposed experiment is needed as a basis for management of human activities as they affect the estuarine environment and, in particular, management of the fisheries for species that may depend on the estuary for recruitment. Particularly interesting will be ascertaining the impact of unusual climatic conditions on survival and recruitment of important commercial species.

The Fluxes of Materials from the Continent to the Ocean

Problem Rivers supply far more dissolved and particulate material to the ocean than do all other sources combined. Yet the composition, character, and reactivity of most of this material is poorly identified. The influence of rivers on the geochemical budgets of many elements remains a major unknown in the cycling of materials through the ocean. Estimates of the composition and rate of fluvial and atmospheric inputs to the ocean can constitute important boundary conditions for chemical budgets in the ocean and can be used to ascertain the impact of human activities in changing the ocean environment. The problem is to determine the net fluxes of particulate and dissolved material (organic and inorganic) to the ocean and to learn how these materials are modified during their transit through estuaries and coastal waters.

Approach (1) Determine the chemical composition and flux of dissolved and particulate loads and the mineralogy of particulate matter in selected major river systems representing different climatic regions. (2) Determine the short-term chemical and biological modification of these dissolved and suspended components as river water mixes with seawater. (3) Determine the fate of particulate and dissolved material in estuaries by using anthropogenic tracers of recent and localized origin and of known behavior (including refractory organic compounds, industrial metals, radioisotopes, and distinctive stable isotopes). (4) Determine the origin (terrestrial, biogenic, volcanic, anthropogenic, or whatever), distribution, and variability of aeolian fluxes to the ocean by atmospheric sampling at appropriate continental and ocean locations.

Local Dynamics Shelf Experiment

Problem Which dynamical processes govern the wind-driven and lower-frequency transient motions over the continental shelf? The problem involves the description and parameterization of the momentum flux (primarily vertical) in and through the surface and bottom boundary layers.

Approach A stable moored array of linear two- and three-axis current meters would be developed and deployed throughout the water column in selected locations to measure both very-high-frequency Reynolds stresses that transport momentum vertically and tidal and subtidal transient motions. Meteorological and bottom-pressure instruments would measure regional surface wind stress and barotropic pressure fields. The experiment would be conducted in both stratified and unstratified conditions, beginning in a broad shelf region where the local circulation is reasonably well known.

Comment A new generation of current sensors—acoustic, electromagnetic, and mechanical—represents a major technological advance in linearity and sensitivity. Completion of the development of such instruments during the next few years will make possible both boundary-layer and interior studies.

OPEN-OCEAN PROBLEMS

Gulf Stream Studies

Problem In its simplest form, the energy supply to mesoscale ocean features can be described as follows. Potential energy is accumulated in the ocean by the action of trade and westerly winds of global scale that drive convergent Ekman transports in the surface layer over the subtropical regions. The downward flux from the surface layer induces a southward displacement of water to conserve potential vorticity and a downward flux of heat that accumulates potential energy above the main thermocline. The ocean response is a basinwide anticyclonic gyre with an intensified western boundary current that converts potential to kinetic energy and transfers potential and kinetic energy to smaller spatial scales.

The boundary currents, such as the Gulf Stream, appear to be unstable, giving up their momentum and energy to barotropic and baroclinic eddies and rings. The eddy field near the Gulf Stream is sufficiently intense to drive a mean recirculation in the surrounding waters, which produces a significant departure from the so-called Sverdrup interior balance. It seems likely that mesoscale eddies constitute the principal mechanisms for transferring energy and momentum to smaller scales for eventual dissipation by friction internally and in boundary layers.

The problem is to determine both the primary and the secondary sources of energy for the mesoscale eddy circulation. Is it possible, for example, that eddies can be generated by instability in regions of strong baroclinic gradients or by features of bottom topography and coastline? Such regions need to be identified and described, and the mechanics of energy conversion must be examined in greater detail.

Approach An example of a large field program that could gather information crucial to the design of a pilot flux experiment is the study of the mesoscale variability of the western boundary region of the western North Atlantic and Gulf Stream. This would be a logical extension of the MODE experiments and other studies of the Gulf Stream system. Because the Gulf Stream is emerging as a major source of mesoscale motions, the mechanics of energy conversion there should be examined in greater detail. Both process-oriented and descriptive experiments should be designed to illuminate the process of energy conversion and strong eddy mean-flow interactions in this energetic system. New instrumental capabilities must be developed to sample the most energetic regions adequately in both time and space.

Comment The velocities of mesoscale-range variability exceed the mean flow over much of the world ocean, making the eddy range dominant in understanding the dynamics of the motion and in predicting it for other applications. Eddies can effect the general ocean transport of heat in a western boundary region in two ways: directly by eddy transport and indirectly by driving rectified flows. The huge deep gyre driven by eddies beneath and offshore of the Gulf Stream is effective in reducing the heat advected poleward by the Gulf Stream itself. Because the thermocline is deeper on the offshore side of the Stream where a substantial portion of the southwestward flow of the gyre occurs, the net effect of the recirculation is to advect heat equatorward. It is clear that changes in recirculation will have major effects on the global heat flux. This unexpected feature of the Gulf Stream requires investigation. It is intimately tied to the dynamics of the eddy regime.

Sverdrup Experiment

Problem It has long been supposed that the dynamics of the central regions of subtropical oceans are governed by a simple linear vorticity relationship. Recently it has been shown that the density field contains enough information to test this idea and to compute features of the flow field that were previously believed to be inaccessible. Since density surfaces in the center of a subtropical gyre are not parallel and do not slope in the same direction, the currents spiral through the main thermocline. This, in combination with a statement of a simple linear vorticity conservation, makes it possible to calculate the absolute field of velocity, both horizontally and vertically. It appears that with a more complete survey of the density field in the center of the gyre, fairly reliable calculations of vertical velocity could be obtained and the role of mixing could be explored. An attempt should be made to obtain a quantitative description of the actual mean circulation in this region of the ocean—a feat so far never accomplished.

Approach In order to make full use of this method, it is desirable to obtain density data in a region measuring approximately 10° of latitude by 20° of longitude and located in the center of a subtropical gyre. A dense and systematic grid of hydrographic stations would be required with measurements each month over approximately two years. Using the emerging technological capability of the new profiling current meters, direct measurements of surface currents should be made within this region and at approximately 10 points surrounding it, in order to obtain a measure of the upper-layer flux of mass and heat and the flux convergence in the region as functions of time. These two independent measurements can be compared through the calculation described above. In addition, measurements by subsurface moored current meters would be welcome for a cross-check of the computed velocities within the main thermocline and in the deeper water. If possible, tracking of deep SOFAR (Sound Fixing and Ranging) floats by autonomous listening stations would be useful.

Comment The midocean subtropical gyres are the simplest and most fundamental elements of the ocean circulation. A full description of their particular dynamics has a natural priority over that of other more limited and complicated regions. The proposed experiment is compatible with other investigations and could provide an instrumented environment for mixed-layer studies, water-mass formation and modification studies (18° water, subtropical salinity-maximum water, Mediterranean outflow water), and gathering of mesoscale statistics.

Transient Tracer Study

Problem As a consequence of technological and manufacturing developments, several transient tracers are now globally distributed. These include tritium, carbon-14, krypton-85, and several fluorocarbon compounds. Key information on the rates of formation of deep water and of penetration of anthropogenic materials into the ocean could be obtained from a program of long-term measurement of the oceanic distribution of such tracers.

Approach Time changes in selected transient tracers should be monitored at certain ocean locations selected on the basis of GEOSECS tritium data and current concepts of deep-water formation. Large-scale experimental releases of certain tracers could provide unique opportunities for diffusion studies.

Comment It has been proposed that the release of CO_2 from the burning of fossil fuels has a significant effect on climate. The proposed program could contribute to an evaluation of the ocean's capacity to absorb CO_2 and other atmospherically transported pollutants. A particularly useful tracer, tritium, has been introduced from nuclear-weapon explosions; because of its short half-life (12 years), there is some urgency in starting the program.

Past Climatic Change

Problem The history of climatic change can be extrapolated back over long periods of time using geological and geophysical methods. The detailed history during the past 5 million years is of particular interest.

Approach (1) Measure detailed profiles of isotopic, chemical, mineralogic, and micropaleontologic changes in long cores that can be dated by paleomagnetic reversals. (2) From these properties, reconstruct changes in features of climatic importance, such as the volume of global ice and the physical characteristics of surface and bottom waters. (3) Obtain cores in which such changes can be monitored in the regions of major water masses and in the transition zones between them. (4) Use spectral and filter techniques to resolve the amplitudes and frequencies of geographic and temporal changes in reconstructed climatic variations. It will be necessary to develop a coring technique that can provide detailed, undisturbed records of the top 100 m of the sediment column and to exploit this technique systematically in a major field program to collect the necessary samples.

Comment Recent work has shown that characteristic amplitudes and periods of climatic change have increased over the past million years. What has caused the general global cooling that culminated in the late Quaternary ice ages? The few piston cores available are inadequate to reconstruct details of the climatic history and, hence, to permit framing or testing of theoretical models. For example, the distribution of volcanic ash must be better known in space and time in order to assess the possibility that changes in the atmospheric ash content have affected climate. Information on the sequence of oceanic events associated with the ending of the last interglacial period should provide insight into the future development of the present climatic regime.

EQUATORIAL DYNAMICS

Equatorial Circulation Dynamics

Problem General problems have been discussed earlier. Specific problems include the coupling between ocean and atmosphere, the transmission of low- and high-frequency signals in the equatorial waveguide, and their interaction with the circulation of the eastern boundary region.

Approach The eastern equatorial Pacific is an appropriate region for this study in that it is accessible, the phenomena and processes appear to be uniquely developed there, and a background of descriptive and theoretical studies already exists. Measurements will be required over a period of years to resolve the interannual variability. The atmospheric velocity field would be determined from ships, aircraft, and satellites; arrays of moored current meters would measure oceanic velocities. Variations in the temperature and density fields would also be monitored.

Comment A preliminary experiment is being organized as part of the First GARP Global Experiment (FGGE).

Equatorial Ecosystem Dynamics

Problem How do the large-scale disturbances that are propagated across long distances at the equator and poleward along the eastern boundaries of the ocean basin affect the local productivity of the highly productive ecosystems present in these regions? What is the interaction between remote climatic forcing and local meteorological forcing, and how does this regulate the productivity of equatorial and coastal regions? What are the processes whereby disturbances arise in the atmosphere, are propagated in the ocean, and lead to biological consequences with global economic impact?

Approach The proposed investigation in the eastern equatorial Pacific would take advantage of the physical investigation described above and the continuing, long-term sampling programs off the coasts of California and Peru. The physical, chemical, and biological methods have been developed in the comparative ecosystem analysis programs of the present decade. This experience suggests some simplification of the measurement program but demonstrates the need for better temporal resolution of both high- and low-frequency variability. Continuously recording fluorometers would be incorporated in the current-meter mooring arrays to obtain similar resolution of the physical forcing and the biological response.

Comment This program is a particularly good example of a basic oceanographic investigation with an immense potential economic impact. The disturbances that arose in the Pacific and that were propagated along the equator to the west coast of South America had devastating economic impacts in 1972 and 1976. Such impacts might be avoided by using management strategies that draw on the agricultural as well as the marine resources of the affected nations. Bringing terrestrial and ocean food-production systems into a predictive relationship with climatic changes would be a major advance in the use of science to improve human conditions.

SOUTHERN OCEAN

Southern Ocean Dynamics

Problem General problems in the Southern Ocean have been discussed earlier. During the past decade, the space and time scales containing the most energy have been identified in a few selected regions of the Antarctic Circumpolar Current. The major dynamical question of the overall momentum and energy balance of the current and the role of bottom topography in establishing this balance is yet to be answered. Other related problems include clarifying the large-scale air–sea interaction in the Antarctic Circumpolar Current system, describing the effect of the annual variation in the extent of sea ice on the heating and mixing of the ocean surface layer and on driving the circumpolar current, and determining the nature and variability of air–sea exchanges and their relation to water-mass formation and the associated meridional circulation.

Approach Arrays of moored and drifting instruments are required to establish the effect of bottom topography on the current. Data collection by satellite and ships of opportunity and the establishment of sea-level measurement stations around the Southern Ocean would contribute to the study of large-scale current variability and large-scale interaction of the Antarctic Circumpolar Current and the southern hemisphere winds. The amount and variability of subantarctic water-mass formation could be studied by experiments west of Chile, south of Australia, and north of the Weddell Sea. Determining anomalies of water temperature and sea-ice extent over the entire pack-ice region would contribute to defining the meridional heat transport in the upper ocean at middle and high latitudes. New technology will be required for making winter measurements, for example, transmitting current meters, disposable moorings, remotely tracked SOFAR floats, and automatically deployed XSTD's. Transient chemical tracers can be used to study long-term variability and mixing rates and processes in the ocean.

Comment The ISOS project, the continuing Antarctic Circumpolar Survey from the *Islas Orcadas* (formerly USNS *Eltanin*), and the International Weddell Sea Expeditions have established space and time scales in selected regions of the Southern Ocean. Successful studies have accumulated substantial information in the Drake Passage and the Scotia Sea; experiments south of Australia began in early 1978. Initial results have revealed previously undocumented mesoscale variability (ring formation and a multicored Antarctic Circumpolar Current). As the data base has increased, studies have begun on the long-term variability of the flow in these regions and the effect of wind forcing on the formation of intermediate waters. Preparation for investigations in the 1980's must await analysis of the full data set including that to be collected during 1978 and 1979.

Southern Ocean Ecosystem Dynamics

Problem How is the large stock of primary consumers (krill) maintained and renewed in the turbulent, light-limited Southern Ocean? What role does the spatial discontinuity of krill and phytoplankton distributions play in the ecological strategy of the region? Is this high-density population so slowly renewed that large-scale and continued harvesting will not be possible? To what degree has the reduction of the principal consumers of krill, the baleen whales, left the ecosystem in imbalance?

Approach Comparative ecosystem analysis based on successful programs in other highly productive ocean ecosystems would be employed to determine the basic structure and function of the low-diversity, short food chains that characterize the region. Physical, chemical, and biological methods establish the key processes that shape the ecosystem and the rate functions regulating the fluxes of material and energy. Intensive studies during the brief austral summer would compare the trophic structure in the highly productive waters of the Weddell Sea with the much less productive waters off the Ross Ice Shelf. Spatial resolution of the physical structure of the euphotic zone in the area of krill patches will reveal the coupling between physical and biological discontinuities. Intensive resolution in time of the physical structure of the upper ocean is required to elucidate the coupling between short-term atmospheric forcing and phytoplankton growth regulation. Detailed analysis of the *in situ* feeding dynamics of krill and of the impact these grazers have on their phytoplankton prey is required.

Comment Exploitation of the krill resources of the Southern Ocean offers one of the last opportunities to expand significantly the quantity of protein that can be obtained from the sea. Because of the immense potential harvest, several nations have designed ships specifically for the purpose of exploiting this international resource. It is clear that there is a high concentration of krill in certain areas, but it is by no means obvious that the Southern Ocean ecosystem has the capacity to maintain these populations in the face of heavy harvesting. The fundamental character of the productivity is not clear; analogies with other productive ecosystems do not provide a sure basis for managing the harvest of this resource in a manner that will ensure steady-state productivity. It is imperative that the analysis of the basic ecosystem be started before it is substantially modified by extensive fishing.

SEAFLOOR SEDIMENTS

Benthic Boundary-Flux Studies

Problem How do the processes of particle formation, transport, and dissolution operate, and what are their variations in time and space? What is the nature and rate of chemical reactions at and fluxes across the ocean-floor boundary, and how do these vary with dynamical and sediment properties and particle input? What are the physical and chemical processes that affect the vertical and horizontal transport of material to and from the deep seafloor?

Approach In selected areas of contrasting ocean-surface properties, such as upwelling, equatorial, high-latitude, and oligotrophic areas, the formation and flux of particles through the water column would be studied with moored sediment traps and large-volume filtration. Particles collected by these methods would be analyzed for inorganic and organic species and for selected radioisotopes. Time-series studies of short and seasonal scales are important. The turbulent-flow field and vertical eddy diffusivity would be characterized within the bottom boundary layer with bottom-moored acoustic, optical, or electromagnetic sensors. Critical erosion stresses would be measured *in situ* for the major deep-sea sediment types. The controlling factors and depths of transport of particulate and dissolved material below the sea bottom would be determined. Sediment traps and nephelometers would measure the distribution and characteristics of suspended material near the bottom, and the gradients and their variability of physical and chemical properties would be defined. Biological studies would involve definition of benthic communities and biological rates. For all elements of the program, *in situ* experimentation and measurement on the ocean bottom would be required. Physical studies defining the variability of bottom-water dynamics and interactions should be included.

Comment Models describing the composition of ocean water require a better understanding of the magnitude and nature of particle fluxes. Fluxes across the ocean–sediment boundary affect both the composition of the ocean and its capacity to respond to changes. Activities such as oil and gas development and production, the dumping of industrial wastes, and deep seabed mining have raised the priority for investigation of the bottom boundary layer.

Geotechnical Properties of Marine Sediments

Problem What are the processes that lead to the postdepositional alteration of the organic composition and clay mineralogy of marine sediments, to their lithification and cementation, and to the development of their acoustic and geotechnical characteristics (those pertaining to mass physical properties)? What factors govern the rates of these processes?

Approach (1) Define the diagenetic processes occurring below the zone of bioturbation by means of geochemical, geotechnical, and acoustic measurements. Such processes include solid-solution reactions that bring about modification of the composition of interstitial waters and recrystallization of unstable sediment components, which cause changes in porosity, permeability, and consolidation of sediments. (2) Relate these processes to the thermal, pressure, and chemical changes arising from increasing age and depth of burial.

Comment Many of the important processes that result in the lithification of unconsolidated sediments occur in the upper few hundred meters of the sediment column. The consequences of these changes are apparent in rocks exposed on land. However, the rate-controlling conditions are poorly understood and can only be determined by measurements in unlithified marine sediments and by coordinated laboratory studies. Improved methods for obtaining long, undisturbed cores of the upper 50 to 100 m of sediment are needed for these studies.

DEEP SEABED

Deep Lithosphere and Asthenosphere Studies

Problem How do the deep oceanic lithosphere and asthenosphere evolve over geological time? How does the evolution constrain models of a possible driving mechanism?

Approach By means of long refraction lines (2000 km), elastic waves would be recorded from depths beneath the upper-mantle low-velocity zone and the transition zone at the olivine–spinel phase change near a depth of 350 km. Data on travel time and amplitude versus distance complemented by synthetic seismograms can yield profiles of velocity with a depth resolution of a few kilometers, in contrast to the resolution of 40–50 km currently possible from surface-wave studies. Stable spectral estimates of the earth's response at periods long enough to resolve properties to a depth of about 40 km would be obtained by deploying passive electromagnetic instruments on the oceanic lithosphere for periods of several months. At each test site, an area approximately 2000 km in diameter would be surveyed to define the surface topography and gravity field.

Comment Recent theoretical and experimental results point to the existence of small-scale convective cells with a wavelength of several hundred kilometers at the base of the lithosphere. High-resolution gravity and topographic data along closely spaced lines at each study area are needed to detect such cells. Even if a convective pattern is not detected, the measurement will constrain models of the rheology of the deep mantle in terms of the response function between gravity and topographic variations.

The relatively small number of recording instruments available for refraction studies (for comparison, many of the European continental lines employ more than 50 seismographs) and the finite shipboard space for storing explosives require the participation of a large number of institutions and their facilities. Because the seismic velocity is likely to be anisotropic, long refraction lines should be run in more than one direction at each locality. The magnitude and vertical extent of such anisotropy constrain models of both mantle petrology and convective flow in the asthenosphere. Complementary continental refraction lines and electromagnetic experiments currently in progress will assist in evaluating the nature of the continent–ocean differences that appear to extend hundreds of kilometers into the mantle. Projected improvement in instrumentation and navigational capability should permit shipboard gravity measurements of sufficient accuracy (1 mgal) to detect small variations in lithosphere–asthenosphere properties.

Seawater–Rock Interactions

Problem What is the nature of the interaction between geothermally heated seawater and basaltic crust? How do the elemental fluxes to and from the crust produced in this process affect the composition of the oceanic crust, seawater, and deep-sea sediments?

Approach Recent dives by the deep submersible *Alvin* on the crest of the Galapagos rift zone have demonstrated the utility of submersibles for collecting samples of hydrothermal solutions resulting from interactions between seawater and oceanic crust. Continued efforts are needed to locate and sample other submarine hot springs in a number of tectonic settings, using surface ships and deep submersibles. These studies should (1) determine relationships between various chemical species and conservative tracers in the hydrothermal solutions; (2) map the distribution of these tracers around the hydrothermal source regions; (3) determine by radiometric dating of sediments the anomalous accumulation rates of highly insoluble species, such as the transition elements, in sediments near hydrothermal sources; and (4) measure the duration of activity and the temporal variability of flow rates as well as the chemistry of several hydrothermal vent systems.

Comment Understanding of the qualitative effects of reactions in the oceanic crust on the composition of seawater has reached the point where a major effort is warranted to design experiments capable of quantifying these effects. Because the location, flow, and composition of the hydrothermal systems may be controlled by the tectonic setting, studies on fast-spreading and slow-spreading ridges and on fracture zones are necessary to evaluate the global geochemical and thermal influence of this process. Instruments such as transponders for precise navigation, CTD's, and submersibles with adequate endurance and capability for sample collection and data logging have developed to the extent that rise-crest hydrothermal systems can be reliably located and sampled.

The study will begin to identify the factors controlling the location and lifetimes of seawater hydrothermal systems. Because these systems concentrate valuable metals, a characterization of the solution by isotopic and compositional studies will enhance the understanding of the processes that formed many currently exploited ore deposits on land.

5 On the Conduct of Cooperative Ocean Research

PROJECT SELECTION AND DEVELOPMENT—GENERAL PRINCIPLES

The basic objective and operating characteristics of the IDOE distinguished it from other programs in the field of ocean science and engineering. As discussed earlier (Chapter 3), the IDOE has been a program of fundamental research that linked the achievement of oceanic knowledge and understanding with the more effective use of the ocean and its resources. We believe that a similar but broadened objective would provide an appropriate focus for the proposed successor program. The following statement reflects this point of view:

The objective of the post-IDOE program should be to achieve the comprehensive knowledge of ocean characteristics and their changes and the profound understanding of oceanic processes required for more effective utilization of the ocean and its resources, protection of the marine environment, and forecasting of weather and climate.

To achieve this objective most effectively, post-IDOE projects should have most, if not all, of the following characteristics:

1. *Scientific quality and significance:* The proposed work must be of superior scientific quality. There should be a high likelihood of producing a significant increase in fundamental knowledge and understanding, and the investigation should yield results of widespread interest.

2. *Cooperation:* IDOE projects have been characterized by cooperation among scientists from different disciplines, institutions, and countries. Cooperative projects have been particularly effective when they have been sharply focused on a central problem of common interest. While it has not been required that each project be interdisciplinary, interinstitutional, and international, these cooperative aspects have been strongly encouraged. On occasion these interactions may have imposed additional costs on the conduct of research; nonetheless it is widely agreed that they have brought great strength to the program and should be continued in the future. The size and complex interactions of the ocean have of themselves always stimulated cooperation in its investigation. Yet, because of the magnitude and duration of the projects, the IDOE and its successor provide a special opportunity for studies that require the diverse capabilities of scientists, wherever they may be employed.

3. *Relation to application:* Fundamental research on oceanic phenomena and processes is required for the long-term solution of societal problems. The post-IDOE program should consist of projects of a fundamental nature rather than of those directly tied to short-term applications, which would be more properly funded by other agencies. However, the relevance of each proposed project to the long-term objective of the program should be identified, and special attention should be given to projects whose results are particularly likely to contribute to the solution of an important social or economic problem.

4. *Magnitude and duration:* It is anticipated that projects of the post-IDOE program will involve a number of Principal Investigators, normally from several institutions, and that they will be conducted over a period of several years. Relevant criteria include the following:

(a) The size, complexity, and duration of the field programs impose special requirements for coordination.

(b) Instruments, equipment, and facilities are large and complex, and their sharing among institutions is appropriate.

(c) Large and expensive equipment must be developed.

(d) Field programs and modeling efforts must be integrated.

(e) The magnitude and complexity of the project requires planning and management procedures like those of the IDOE.

(f) Stable funding over a prolonged period is necessary.

The magnitude and duration of the investigations will vary from project to project, and precise criteria should not be set. The funding level will reflect the number of investigators needed as well as unique requirements for engineering and logistic support and for equipment development.

Projects will normally continue for several years. More than a year is usually required for proposal development and review. A similar period is consumed in organizing the research team and in developing and testing research equipment and procedures. Field observations and experiments will require two or three years, and an additional year or more may be necessary for data analysis and interpretation. A major project may require at least five years for its full implementation, with variable annual costs depending on equipment needs and the scheduling of field operations.

Projects of intermediate size, larger than those normally funded by the Oceanography Section of the National Science Foundation but smaller than the normal IDOE project, may also arise. Such projects may continue for only a year or two, may then be terminated, or may subsequently develop into major projects. Intermediate-scale and pilot activities represent an important gap in the range of project size, and their funding should be included in the post-IDOE program.

Ensuring scientific quality is, in part, a function of the process of proposal development and selection. During the IDOE, the procedure for project development has commonly been that, following preliminary discussions and correspondence to establish the interests of scientists and of NSF/IDOE in a certain topic, a workshop was convened to establish the state of scientific knowledge of the topic and to identify key problems requiring investigation. Participants then selected scientific leaders responsible for design of the project, and an announcement, broadly distributed to the scientific community, invited submission of letter proposals. From the total set of such proposals, the Steering Committee selected those contributing to a common approach, and a collective proposal was developed and submitted to the NSF.

A scientific advisory committee has been responsible for the design of the project, including outline planning, specification of measurements, and determination of technology, personnel, and funding needs. Such committees have included some of the Principal Investigators and other interested and knowledgeable people with no direct involvement in the project.

While these procedures have generally been regarded as satisfactory, they have occasionally been criticized as excluding scientists who are qualified and interested in participation. The staff of the IDOE has made strong efforts to guard against such exclusion. The following suggestions are intended to ensure that the process of project development and selection is open to all concerned.

For major projects, it is essential that interested scientists with relevant experience and ideas be impartially included in project development. This will require early and widespread notification of opportunities for participation in the initial stages of planning. Depending on the nature of the project, scientists from both large and small institutions and engineers and other potential users of research results might be involved in project development. Inclusion of representatives of government agencies with related interests could aid coordination with appropriate projects under their sponsorship.

It is important that the development of projects be a flexible process so that unnecessary delays, costs, and frustrations can be avoided. A variety of approaches should be encouraged. For example, a group of interested scientists might cooperate in developing a project and submitting a collective proposal. It should be possible to incorporate other participants with complementary interests and skills. Some projects may be so well defined as to make preliminary workshops unnecessary. Where projects are unusually complex, planning grants may be required. In some cases, it may be desirable to fund pilot studies before a project is fully developed.

The initial screening of appropriate topics is a critical stage and should be broadened beyond consideration by the relevant program managers. For example, involvement of an advisory board for the Division of Ocean Sciences as a whole might be appropriate and would ensure coordination with related activities of other NSF programs.

Once a proposal has been received, IDOE procedure has been to send it and its individual components to a number of scientists for mail review. Special panels of experts have then been convened to consider the mail reviews and to recommend for or against support. An IDOE Proposal Review Panel has considered all proposals that are not reviewed by special panels, received the reports of those panels, and recommended action to be taken.

This process of review is generally regarded as fair and effective, and there are few suggestions for its improvement. Experts from other fields of science and engineering, as well as representatives of other agencies, could contribute to the review process. Review of collective proposals must accommodate the fact that certain of the supporting components, even though of high quality, are only viable as parts of the whole; this is particularly likely in the case of interdisciplinary projects. At the level of the program Proposal Review Panel, where there is an awareness of the content of other projects, opportunities should be sought to coordinate overlapping projects, particularly where there are common needs for equipment development.

PROGRAM AND PROJECT MANAGEMENT

During the IDOE, program managers in NSF have overseen project management organized by the scientists and institutions concerned. The NSF staff involved with the IDOE has been kept at a reasonable size and has discharged its oversight functions effectively. However, the same division of topics (Environmental Quality, Environmental Forecasting, Seabed Assessment, and Living Resources) by which the IDOE office has been organized is not likely to be appropriate for the future program, which will not fall so conveniently into predetermined packages. Projects concerned with related scientific problems should be grouped together under individual program managers. It is important that these officials be familiar with logistic and operational requirements as well as with the scientific problems addressed, that projects be allocated among them equitably, and that, to the extent possible, a program manager remain with a project throughout its life.

The typical organizational arrangement for project management has consisted of a steering (executive) committee assisted by a scientific advisory committee, a project manager, and various expert panels. The key element has been management by the participating scientists. This approach has proved highly responsive to scientific needs. While critics have alleged instances of apparently inefficient and wasteful use of scientific talent and have claimed that professional project managers would be more effective, most scientists with IDOE experience agree that the benefits of management by scientists far outweigh the lost time that they would otherwise devote to their research.

As will be emphasized in the discussion of personnel requirements, competent project managers are difficult to find, and their contribution to the success of projects must be properly recognized.

While some portion of project funds must be allocated to administrative costs, participating scientists are anxious to keep these costs to a minimum. Thus, the organizational and administrative structure of future projects should be kept as simple as possible. As circumstances vary from project to project, so will details of the management structure. However it is arranged, it must encourage individual efforts while integrating them into the rest of the project, and it must accommodate new ideas as they arise.

After the NSF has decided to support the proposal and has determined the funding level, detailed planning becomes the responsibility of an executive (or steering) committee and the project coordinator, who works with the scientific advisory committee. Success of the arrangement depends on close interaction with NSF/IDOE and the Principal Investigators. Activities of the executive and scientific advisory committees, which must have the full confidence of participating investigators, are subject to review by the NSF.

While the success of management by participating scientists in IDOE projects has generally been outstanding, there have been important differences from project to project. Organization of the post-IDOE program could benefit from an understanding of the reasons for this variability in success. The IDOE Marine Affairs Program has provided an opportunity for research into these differences; its continuation would permit further support for such work.

The long-term nature of projects must be recognized in the proposal and review process. With proper initial review and selection of Principal Investigators, projects should not require annual rejustification. At the same time, rigorous periodic review and updating of objectives and approaches are necessary to ensure the maintenance of high standards of scientific quality. The frequency and timing of these reviews are critical. While they must provide accurate evaluation of the quality and progress of the project, they must not be so frequent or erratically scheduled as to disrupt scientific productivity. Projects can be structured into logical work and time segments, and the review schedule should be established during project development and should take advantage of the natural plateaus in achievement and understanding that can be expected.

Total duration should also be specified during project development so that participating scientists and the NSF are able to deploy their resources in a deliberate and rational manner. Large projects can develop considerable inertia because of the multitude of investigators involved and the substantial investment in facilities. It is sometimes difficult to terminate them and replace them with more fruitful efforts. The inertia can be overcome by scrupulous review at successive stages of project implementation and by the early establishment of well-defined termination points.

INTERNATIONAL DIMENSIONS OF COOPERATIVE MARINE RESEARCH

From its inception as a recognizable branch of scientific inquiry some hundred years ago, oceanography has been seen as an international science. The reasons for this are many. They include the following:

1. The world ocean is an interconnected and interacting continuum.
2. There is free exchange of scientific information among oceanographers in different countries.
3. Through international cooperation in research the coverage of exploration can be extended and the study of processes can be more intensive.
4. With recent extensions of national jurisdiction and the forthcoming new legal regime, coastal states will control research within 200 miles of their shores.

Until recently, scientific collaboration in marine research largely had developed among a few industrialized countries possessing comparable scientific capabilities and common interests. Data and scientific findings have been exchanged freely; scientists have participated in research at foreign institutions and aboard foreign ships; joint expeditions have been organized to pool resources of intellect and material in a common search for knowledge. Cooperation between scientists in the United States and those in developing countries has progressed more slowly but has accelerated in recent years. One of the most successful cooperative efforts has been with countries of western South America in the IDOE/CUEA project.

The international character of the IDOE was stressed in its title and in much of its planning. International participation was encouraged; the *IDOE Progress Report,* Volume 6 (1977) lists 47 countries that took part in the IDOE. Yet the extent of international cooperation in the program has been somewhat disappointing and certainly has fallen short of that originally envisioned. Few countries allotted additional funds for IDOE programs, and most of the international participation has been that of foreign scientists on projects that were carried out by the United States. Perhaps the early support for the IDOE in the IOC and the United Nations led to unrealistic expectations about the pace at which meaningful cooperation in ocean science could be developed.

What are the implications of these considerations for the proposed new program? Although the program need not be described explicitly as "international," its products will be of international interest, and its implementation will require international cooperation. That cooperation will continue to be highly developed with laboratories in industrialized countries, within waters under their or our jurisdiction, or on the high seas. Cooperation with scientists of developing countries will be fostered by mutual interest in scientific questions and by the desire to assist these countries in becoming capable of participating in, and utilizing the results of, marine research.

International cooperation will also be affected by the Law of the Sea. Since the 1958 Convention on the Continental Shelf, seafloor research by foreigners on the shelf has required the consent of the coastal state. This consent requirement is being enlarged, both by unilateral extensions of jurisdiction and by the emerging consensus of the United Nations Conference on the Law of the Sea (UNCLOS) to require consent for all research within 200 miles of the shore. In addition to requiring consent, the UNCLOS text would impose a system of advance notification and an extensive set of obligations. These include the right of participation; provision of reports; access to data and samples; assistance in assessing data, samples, and results; and provision of information on program changes. An intent of these obligations is to ensure that the coastal state can share in the benefits of research.

What are the consequences of coastal state control over research for the post-IDOE program? Although some problems will arise with the industrialized countries, the major impact is likely to be on investigations off the coasts of developing countries. These projects will have to be cooperative; scientists from the coastal states may participate from the initial planning to the data analysis and interpretation stages. Some projects are likely to be of particular interest to these countries, for example, studies of coastal and estuarine processes, river inputs, and other chemical fluxes. Short-term applications will likely interest developing countries more than fundamental discoveries, and cruise plans may have to reflect these interests.

Inasmuch as political considerations and bureaucratic delays might affect access, the planning of distant water operations will become more uncertain. Planning delays and the participation obligation will make projects more costly. Furthermore, assisting countries with the utilization of scientific findings may be expensive. There will be even larger requirements for technical assistance expenditures. Nevertheless, these costs should eventually be outweighed by the benefits of an expanded community of informed scientists and of a broader concern for the wise use of the ocean and its resources.

It must be emphasized that, unless additional funds are provided, either in the research budget or in a separate allocation, these additional expenses will result in correspondingly less research. A separate allocation is preferable for the technical assistance aspects, and it is proposed that the NSF take the lead in negotiating suitable arrangements with the Department of State, with Sea Grant, and with other appropriate federal agencies.

International cooperation in ocean research is facilitated by a variety of mechanisms, informal and *ad hoc,* or formal and institutional. In some instances, intergovernmental bodies such as the Intergovernmental Oceanographic Commission have been able to assist in arranging access to foreign zones. In others, international organizations have been instrumental in the organization and coordination of cooperative investigations. Experience has shown the greatest success when arrangements have been relatively informal. For example, when laboratory-to-laboratory agreements are effective, it is undesirable to invoke intergovernmental machinery. The NSF should continue to monitor international arrangements and should support those that contribute to the accomplishment of scientific programs.

INSTITUTIONAL AND INTERAGENCY INTERACTION

The post-IDOE program will involve interactions among academic institutions cooperating in the development and implementation of individual projects, the various elements of NSF concerned with the support of ocean research, several agencies of the federal government, and industry.

Traditionally, research projects have been developed, proposed, and implemented by scientists within a single institution, which has the responsibility for meeting conditions of the grant or contract. In contrast, a unique feature of the IDOE has been the collective development and implementation of projects by groups of academic institutions. The usual mechanism for this interaction has been the steering or executive committee with its scientific advisory committee. Both committees have consisted of scientists from participating institutions. Collective proposals have been prepared and submitted by these bodies on behalf of the several institutions involved. While a single institution may undertake administrative responsibilities for the entire project, individual grants are made to the participating institutions to cover their components of the project.

In the evolution of this process, two problems were anticipated. First, the collaborative approach was thought to be incompatible with the customary competition among institutions for their share of the limited funds available. Second, it was feared that loyalties to the parent institution would disappear and that control by the institution over its research activities would be weakened. Neither problem has proved to be serious. The competitive position of institutions participating in IDOE projects has, if anything, been strengthened, inasmuch as these institutions have benefited from the funding stability offered by long-term projects. Furthermore, there is no evidence that institutional loyalties have been weakened or that institutions have been committed to research activities against their will. Support among IDOE scientists for the collaborative approach has been widespread.

Judging by the scientific opportunities identified early in this report, future projects are likely to be at least as complex as those in the IDOE, and the collection of talents required for attacking these problems will seldom be available within a single institution. Furthermore, future projects will continue to require the pooling of equipment and facilities among institutions. Thus it is highly desirable that interinstitutional links such as those employed in the IDOE continue to be an attribute of the post-IDOE program.

The IDOE program has represented somewhat less than one quarter of the total ocean research program of the National Science Foundation. In fiscal year 1977, NSF marine activities included: IDOE ($17.1 million), Ocean

Science ($17.7 million), Oceanographic Facilities and Support ($18.4 million), and Ocean Sediment Coring Program ($12.8 million). The Office of Polar Programs provided $6.9 million for marine science; some $2.2 million of related activities was supported elsewhere in the Foundation.

There is a steady flow of ideas and activities among these NSF programs. The Oceanography Section normally funds individual scientists, and it is within this program that many of the fundamental and innovative ideas are developed. Projects of the IDOE have often arisen from research of this sort and represent further development and integration of such ideas. While projects of the Oceanography Section usually fall within a single discipline, the large IDOE projects provide the principal opportunity within the NSF for interdisciplinary approaches. Oceanographic facilities and support for both programs are provided by the appropriate NSF office. Projects in polar waters are funded both in the IDOE (e.g., International Southern Ocean Studies) and in Polar Programs (e.g., Processes and Resources of the Bering Sea Shelf).

The IDOE has not been the only large marine-science program that has operated during the 1970's, nor does it seem likely that the proposed successor program will stand alone. The Deep Sea Drilling Program (DSDP) and the International Program of Ocean Drilling (IPOD) are conspicuous examples of programs that, while resembling the IDOE in many of their attributes, have been separately funded and managed within the NSF. Research supported by the IDOE and the NSF Oceanography Section has contributed to the selection of drilling sites and has made possible the analysis of DSDP/IPOD collections; findings of the drilling programs have inspired investigations funded by other programs. Future drilling will depend more than ever on subsidiary regional information that can result from the efforts of post-IDOE and other investigations. It is anticipated that the fertile interaction among oceanographic programs, both large and small and relating to drilling and other problems, will continue and grow during the post-IDOE period.

Two conclusions can be drawn from these considerations. First, since a program of large-scale studies like those pursued in the IDOE depends on the ideas, methods, and knowledge developed by individual investigators, growth in support for large-scale studies should not be at the expense of research by individuals. Second, the mutual interdependence of ocean research projects in all the various offices of the NSF requires an effective mechanism for their coordination.

Recognition of this interdependence has led to suggestions that several of these programs should be merged. We believe, however, that there are clear-cut advantages in keeping the successor program to the IDOE as an identifiable entity. Both the processes of development, review, and selection of the larger cooperative projects and their management and administration call for different approaches than do the smaller and less complex studies of individual scientists. Thus, rather than combining projects of all sizes within a single administrative unit, it seems better to continue to use the strengths and diversity of the various NSF ocean-research programs, while, at the same time, ensuring their integration through an appropriate coordination mechanism.

Interaction among the several agencies of the federal government concerned with ocean research should be viewed in the context of the total Federal Ocean Program, which in fiscal year 1977 amounted to approximately $1 billion ($943.3 million). Of this, about 15 percent ($146.4 million) was defined as ocean research funding, of which the NSF spends about one half. The rest of the ocean research budget is spent by other agencies, including the Department of Defense ($33.1 million), the Department of Commerce ($23.5 million), and the Department of Energy ($13.1 million). The Department of the Interior sponsors important programs not usually defined in federal budget breakdowns as "ocean research." The USGS spent an estimated $39.7 million for research on the continental shelf in fiscal year 1977, while the Bureau of Land Management spent an estimated $56.6 million on environmental studies related to outer-continental-shelf oil and gas development. In the latter program, a significant amount of work was done by academic institutions under contract or subcontract.

Several IDOE projects have featured cooperation among funding and operating agencies. Perhaps the most extensive such interaction has been in the North Pacific Experiment (NORPAX), which has been funded jointly by NSF/IDOE and the Office of Naval Research (ONR), and in which several elements of the National Oceanic and Atmospheric Administration (NOAA) and other bodies (e.g., the Inter-American Tropical Tuna Commission) have also participated. In this and other projects (e.g., MODE and POLYMODE), ONR is the only agency that has joined NSF/IDOE in funding IDOE investigators in academic institutions.

Other agencies, notably NOAA, participate through coordinated operations when a project is perceived to be important to their missions. Interagency cooperation has also included the exchange of information and, on occasion, the joint use of facilities. Mission-oriented agencies should invest in fundamental research since it provides an essential basis for their solution of short-term problems. Those agencies with in-house laboratories tend to conduct such research themselves rather than contribute to the support of academic research. Although we believe that mission-oriented agencies should contribute directly to the support of fundamental research projects at academic institutions, in practice this is seldom achieved. Agencies have widely varying funding practices, and even if joint sponsorship of projects can be envisaged, long lead-times and bureaucratic complications seem to be inevitable. The problem is aggravated at present by the absence of a means of coordinating the ocean program of federal agencies.

The necessity for interagency cooperation becomes particularly evident as emphasis on research in estuaries and in coastal waters increases. There is a clear need for joint planning among agencies concerned, together with agreement on the allocation of responsibilities in programs of common interest. Mission-oriented agencies could agree to undertake subtasks contributing to the achievement of the scientific goals of such programs; even if financial support cannot be provided for academic research, ships and other support facilities might be made available.

As noted elsewhere, progress on some problems is likely to be impeded by the inadequacy of available technology, and if the post-IDOE program is to succeed, new means must be sought to further the development of such technology. Several agencies (e.g., Navy Department and NASA) have important engineering capabilities. Organizations such as the Navy Department and NOAA need some of the same equipment and facilities required by academic research. Therefore, where appropriate, such agencies should join in the development of essential major items of equipment, either through joint funding of this development or through assignment of their personnel and facilities to the task.

PERSONNEL REQUIREMENTS

The conduct of large scientific projects such as those developed under the IDOE requires personnel in the categories of scientific leadership, conduct of research, operation of equipment and facilities, and management and administration. Those in the first two categories normally are specialists in some aspect of marine science, and their availability could well affect the level of activity in future programs. The training of specialists is time consuming (a PhD in oceanography usually requires at least four years of graduate study), and it can be argued that oceanographic programs can only grow as the pool of competent scientific manpower increases.

The most serious shortage appears to be in the category of scientific leadership. Experienced and imaginative scientists capable of developing and carrying out major projects have always been rare, and they cannot provide the required leadership for an endless series of projects. The IDOE and other programs of large-scale marine research provide an invaluable training ground for the identification and development of potential leaders.

The conduct of research involves midlevel and junior scientists, graduate students, research assistants, and technicians. Expansion of graduate instruction programs in oceanography during the last few decades and immigration into oceanography of scientists from other fields has until now provided adequate numbers of such personnel. Although the availability of specialists varies somewhat from field to field, existing educational programs may be adequate to provide the necessary scientific personnel; the Manpower Panel of the Ocean Sciences Board is now studying this question.

Graduate students are essential members of research groups. Most investigators believe that IDOE projects have greatly expanded the interdisciplinary knowledge and experience of students and that funding for their participation should be continued and strengthened in the future program.

Research assistants and technicians require much less specialized training, and the supply appears to respond adequately to demand. The problem is to develop and maintain skilled research support groups in the face of significant fluctuations in project funding. In some cases, after the original tasks have been completed, the capabilities of such groups should be made available to subsequent projects. Similarly, although suitable technical support staff generally can be assembled for the design, construction, and operation of major equipment facilities, this staff is vulnerable to the vagaries of funding. It is often difficult for academic laboratories to offer skilled personnel adequate salaries comparable with those of industry. The costs of competitive salaries must be recognized in planning for equipment and facility development.

Management of IDOE projects has presented special problems. Good project managers, who should be trained scientists capable of handling logistic and administrative aspects of major operations, are hard to find. The established reward system of universities frequently fails to recognize the value of such activities, and these individuals either go unrewarded for their services or eventually seek more rewarding posts. The assignment and support of talented managers is an essential element in the success of the proposed program.

FACILITY AND EQUIPMENT REQUIREMENTS

RESEARCH PLATFORMS

Most ocean research is done from ships, and this will likely continue through the 1980's. Moored instrument systems are increasingly important but must be deployed and serviced from ships. The capabilities of remote sensing from aircraft and satellites are receiving growing appreciation and will be utilized more in the next decade, but their vision is limited to the surface, and "surface truth" must be supplied from ship and buoy observations. No existing devices can match the research vessel in mobility, equipment handling, and capacity to carry both people and equipment.

Research platforms available to oceanographers include full-time research vessels of the academic and government fleets and the occasionally used chartered private vessels and ships of opportunity. The academic fleet consists of some 30 vessels from 65 to 245 ft (20 to 75 m) in length. Many of the ships need to be upgraded, and a few already require replacement. Academic scientists also have access to government vessels, primarily those of NOAA. The use of such vessels has until now been modest. Use of chartered vessels by academic scientists is also limited, and ships of opportunity are used primarily for monitoring.

A plan for upgrading and replacing vessels of the academic fleet has been developed by the University National Oceanographic Laboratory System (UNOLS). This plan takes into consideration special requirements of the post-IDOE program such as the following:

1. Although there is general agreement that coastal research should be included in the post-IDOE program, large, stable vessels with long endurance will continue to be needed for "blue water" projects. Furthermore, much of the proposed coastal work cannot be done on the small vessels ordinarily considered adequate for such purposes. Coastal scientific problems are often more complex than those in the open ocean, and study of them will require the use of bulky equipment and large scientific parties, particularly when an interdisciplinary approach is taken. Thus capacity and equipment-handling requirements may be as great as those for deep-water projects. On the other hand, range and the consumption of fuel, water, and food are likely to be significantly smaller.
2. The suite of essential shipboard equipment has become more complex and expensive. Modern research vessels require precise navigation (satellite, Omega, radar, for example), versatile communication systems, precision depth recorders, computer-based data-acquisition systems, and versatile equipment-handling gear. Obsolete and poorly functioning equipment can impede or even prevent accomplishment of scientific missions.
3. Some proposed projects require specialized ships. For example, while Coast Guard icebreakers have some limited capacity for scientific work, intensive research in either the Arctic Basin or Antarctic waters will require ice-strengthened research vessels not now available in this country. Similarly, some specialized geological programs require drilling ships such as *Glomar Challenger.*

4. For other projects, dedication of specific ships for periods of one or two years may lower costs and increase efficiency. This approach was taken with *Knorr* and *Melville* during GEOSECS. With the present size of the academic fleet, dedicated ships are most effectively assigned to large, multi-institutional projects.

5. At present, *Alvin* is the only deep [13,000 ft (4000 m)] submarine available for academic research. It has been used for important work in the present decade and is expected to be in continued use during the 1980's. The tender vessel *Lulu,* however, is inadequate and must be replaced with a ship capable of prolonged and distant operation. Many scientists believe that an additional submersible, capable of working at greater depths [to 20,000 ft (6000 m)] and with greater scientific capacity, will soon be required.

6. Some proposed projects would benefit from more exotic research platforms. For example, rapid sampling could be facilitated by speedy surface-effect ships. Significant improvements might be made in general-purpose research vessels, for example using semisubmersible hulls. Such vessels are being developed by the Navy for military purposes. When their performance, reliability, and cost have been established, the desirability of their incorporation in the academic fleet should be considered.

The use of platforms other than academic research vessels is growing. Most of the larger research vessels are in the government fleet, and the possibility of increased use of these expensive resources in the post-IDOE program should be explored. This practice would maximize their utilization and might obviate the near-term need to construct additional large vessels for academic research. The use of merchant ships of opportunity for monitoring the ocean surface layer has already been demonstrated in NORPAX, and an expansion of this coverage is proposed as part of an equatorial Pacific research program during the First GARP Global Experiment (FGGE). Such monitoring is essential to studies of long-term climatic time scales. Offshore fixed platforms and island stations will also be used to extend observations in space and time. Aircraft are being increasingly used for the deployment of expendable bathythermographs (AXBT's) and buoys.

The utility of remote sensing from aircraft and satellites for measurement of the ocean surface has been demonstrated in the present decade. No other approach is capable of providing such spatial continuity and repeated global coverage of observations. The versatility and sensitivity of sensors and the operational availability of ocean surface data are steadily being improved. Although remote sensing will not replace the need for measurements from research vessels, it is certain to make a major contribution to oceanographic programs of the 1980's.

EQUIPMENT REQUIREMENTS

During the IDOE, some major items of equipment were developed for use at sea. This was accomplished within the framework of individual projects; in some cases, the equipment was made available to other projects. For the future program, scientific problems already identified can only be investigated with measurement and experimental capabilities not now available to the academic institutions.

Academic research has not had access to some special equipment and instruments that are used by government or industry. In some cases, there are proprietary (or classification) restrictions, while in others the cost exceeds available funds. Some examples include multichannel seismic systems, multibeam sounding systems, stable platforms, and high-data-rate communication facilities. The availability and cost of such devices must be considered in the budgets and plans for relevant future projects. In some instances, it may be appropriate for the capability to be made available on specific government or academic research vessels for shared use by scientists from several institutions.

Some problems will require the development of major items of equipment. Such equipment needs may be common to several projects. The following examples illustrate some of the required capabilities:

1. Progress in historical oceanography (paleoclimatology) is limited by the inability to recover undisturbed sediment cores from the upper 20–100 m of the seafloor. This zone falls between the capabilities of piston corers and deep-sea drilling equipment; the time period represented is from 0.5 million to 100 million years, depending on sedimentation rates.
2. Important advances were made during the IDOE in moored systems of current meters and other sensors in the deep sea. Yet the study of important circulation systems, such as the Gulf Stream, Kuroshio, or the Equatorial Undercurrent, will remain difficult until these techniques have been modified to survive in the strong shear flows present in these currents.
3. Although new biological sampling devices have become available in recent years, for example, the Longhurst-Hardy plankton sampler and the rectangular midwater trawl (RMT), satisfactory methods are still not available for direct observation of standing stocks or production rates of most pelagic and benthic animals. These methods may involve acoustic or other physical measurements not requiring capture of large organisms. Where smaller organisms such as plankton are collected, instruments are needed for their sorting and identification.
4. While ships of opportunity are essential for extended and repeated

monitoring of subsurface ocean characteristics, few simple instruments like the XBT exist that can be used in a routine way by unskilled personnel.

5. Some proposed shipboard experimental work in biology and chemistry requires isolation from ship motion. This problem might be solved through use of gimballed and gyrostabilized modules in which complex and delicate instruments can be mounted.

6. Networks of seabed-mounted instrument packages for monitoring seismic events, measuring chemical and particle fluxes and near-bottom currents, or conducting other types of *in situ* experiments must be designed, deployed, and maintained. Deployment and maintenance may require the use of a manned deep submersible.

The post-IDOE program should have a major role in the organization and funding of such developments where they are required to meet unique program needs. It will often be necessary to go beyond the capabilities of oceanographic laboratories and to utilize the engineering talent available in universities, government, and industry. Specialized needs arising from stated scientific requirements should be identified early in the program, and engineers should work in consort with the ultimate scientific users to specify, design, fabricate, and test the new devices. In some cases, this can best be done in projects specifically intended for development of new capabilities. Once the equipment was perfected, such projects would be replaced by scientific projects in which the new capability would be applied. Where instruments are potentially usable by other agencies, such as NOAA and the Navy, it would be proper for those agencies to share the development costs. The products of these enterprises should be available to scientists in a variety of appropriate projects. Programs of the scale of the IDOE can leave an important heritage of improved research capabilities for use by later generations of investigators.

Three other points should be stressed concerning equipment needs of the post-IDOE program.

1. Recent advances in microprocessor technology and small computer systems have yet to be fully incorporated into oceanographic measurement systems, for example, into instrument automation, experimental control, and rapid data acquisition and processing, as well as into simulation and modeling of oceanic processes. Most investigators need greater access to adequate engineering and programming assistance.

2. There has been no systematic program for upgrading and replacing obsolete equipment used by marine scientists. Consequently, progress is slowed and research results are less definitive than they should be. Use charges, whereby a replacement fund could be accumulated, are not nor-

mally allowed. An alternative is to provide specifically in project budgets for acquiring "state-of-the-art" equipment.

3. Existing instruments and methods are often insufficiently reliable, precise, or accurate for use in studies of long-term variability. If its findings are to have enduring value, the post-IDOE program must encourage the calibration and intercomparison of instruments and methods and the use of suitable standards.

4. Some modeling efforts of IDOE projects have required the use of large computers such as that at the National Center for Atmospheric Research. Limitations in the availability of time on such machines for oceanographic research may restrict future modeling programs. Computer requirements for the post-IDOE program remain to be developed.

DATA MANAGEMENT AND INFORMATION EXCHANGE

Modern scientific instruments are capable of producing vast quantities of data. The complexities of oceanic processes engender a tendency to oversample, and often many of the data collected are redundant. The experimental nature of advanced research calls into play many novel instruments, and the data they produce can be unusual and difficult to accommodate in central systems set up to handle data of a more classical character. As has been noted, calibration and intercomparison of these novel data are of great importance if project results are to be available and useful to other investigators.

The flood of data, even when of uniformly high quality, must be accommodated in a systematic way through a data-management system. Such systems are commonly organized within individual large projects (e.g., CLIMAP, MODE, NORPAX, GEOSECS), particularly when many organizations are involved in a tightly coordinated program. Data management is more highly developed in large-scale meteorological experiments (e.g., GATE, FGGE) than it has been in oceanography, in large part because of the "real-time" distribution and utilization requirements of such projects. Primary use of oceanographic data, on the other hand, tends to be by the original investigators, after which they are archived for possible later use by others.

The paucity of well-developed data-management plans in oceanography in the past reflects the single investigator/institution focus of programs. The interinstitutional character of many IDOE projects has undoubtedly resulted in great improvement in this area. This progress must be accelerated during the future program, so that the investment in data acquisition will be matched by commensurate efforts to maximize utilization of this information. The government agency responsible for oceanographic data management is the Environmental Data Service (EDS) of NOAA. This agency has worked closely with the NSF in handling the data problems of the IDOE. Participants are required to submit reports of observations/samples collected by oceanographic programs (ROSCOP) to the National Oceanographic Data Center (NODC) of EDS. In some cases, the data themselves are deposited in NODC as log sheets, analog records, punched cards, or magnetic tape; in other cases, they are held by the observing institution and their availability is made known to other investigators.

While NOAA/EDS and NSF/IDOE are doing their best to ensure prompt recording and archiving of IDOE data, use of this information by other investigators often requires an enormous effort of pulling together, normalizing, and evaluating information of varying quality from a number of disparate sources. Whereas the first step is certainly to ensure that all data are centrally archived or recorded, the essential task of making possible full and ready access and utilization must not be forgotten. A mechanism is needed to promote vigorous and continuing interactions among the producers and users of oceanographic data and the entities responsible for their storage and retrieval.

Research results as well as data must be made available to other users in the scientific community, in government, and in industry. The present system of publications, workshops, and symposia is reasonably effective for exchanging information within the ocean-science community. However, it is most effective within individual disciplines. A greater effort is required to ensure that results in one field are known and understood by those scientists working on other aspects of a common problem. Two ways to enhance such communications are the publication of review papers in widely read journals (e.g., *Science*) and the presentation of applicable papers at multidisciplinary scientific meetings.

At present, communication of the results of scientific research to the engineering community and to other potential users is much less effective. Here the problem is one of identification and translation as well as one of dissemination. Several approaches for improving communication have been proposed. Some are considered below.

Where proposed research has particularly direct links to application, representatives of appropriate government agencies and industry might be involved at an early stage in project development and planning. In this way, they will be aware of the implications of the research for their activities and, in some cases, will be able to suggest ways whereby the research might better relate to their problems.

The NSF should undertake a continuing review of project results in order to identify those of potential applicability and to translate these results into language readily understandable to users.

Findings of applied interest should be promulgated by publication of reports using appropriate media and by their presentation to representatives of concerned industries and government agencies.

The lay public is another important constituency, especially as it is called upon to provide continuing support for ocean research. The NSF and the scientists and institutions that it supports must strengthen their efforts to communicate to the public scientifically sound and understandable discussions of ocean problems and research on issues of broad public importance.

The problem of technological transfer is poorly understood; the measures proposed are far from adequate. The IDOE has recently established a Marine Science Affairs Program as a means of relating scientific findings and their social, economic, and political implications. Research sponsored by that program may reveal new ways to facilitate the desired transfer and application of findings. Examples of appropriate research topics include the following:

1. Implications of new knowledge in the marine sciences;
2. National and international contexts within which marine research takes place;
3. Transfer and application of marine-research findings; and
4. Strengths and weaknesses of large-scale research programs such as the IDOE.

FINANCIAL REQUIREMENTS

We have not attempted to suggest the proper priority for the support of ocean research among the complex of items embraced by the national research budget, nor have we proposed the fraction of the ocean research budget that should be allocated to cooperative research of the sort conducted during the IDOE. But we have estimated an appropriate level of funding for the proposed post-IDOE program, based on several considerations discussed in previous sections of this report.

1. A higher national priority for ocean research should arise from the growing potential for conflict among uses and users of the ocean and the urgency for a stronger scientific basis for ocean-policy decisions.
2. The interdependence of research activities at scales ranging from those of the single investigator to those of multi-institutional projects requires

that support for ocean research grow at the level of individual investigators as well as that of the larger projects.

3. Additional support is needed for projects of intermediate scale, involving several investigators but of a lesser magnitude and duration than the major IDOE projects.

4. Upgrading and replacement of elements of the academic fleet will require significant funding above that necessary for fleet operations during the coming decade.

5. Inadequate equipment is beginning to impede progress in some fields, and investment in the replacement of obsolete equipment and in the development, testing, and construction of new devices represents an extra cost.

6. Extensions of national jurisdiction and the new Law of the Sea will increase the cost of conducting research in foreign waters.

Our estimate of funding requirements for the post-IDOE program has as its starting point an estimated 1980 funding level for the IDOE of $25 million. This level is used for the basic program in 1981. An annual growth rate of 3 percent is applied. Additional support for intermediate-scale and pilot projects is estimated at 10 percent of the basic program. Ship operating expenses (not included in recent IDOE budgets) are estimated at 40 percent of the total research operating expenditures (basic plus intermediate and pilot projects). An initial amount of $2.5 million is proposed for equipment replacement, development, testing, and construction. A fixed sum is included as the estimated additional annual cost of research in foreign waters.

The estimate for ship replacement was obtained as follows. A recent UNOLS study has addressed the problem of upgrading and replacing various elements of the academic fleet as they reach retirement age. To maintain the present capability, it recommends that the replacement investment during the next 15 years should be $48 million for major vessels and $46 million for intermediate and smaller vessels. If the present vessels are to be operated and maintained effectively throughout their service life, an additional $1 million per year is needed for midlife refits. Thus an average annual expenditure of about $7.25 million is required. The IDOE share of ship use is about 40 percent; at this rate, the post-IDOE share of ship replacement and refit is about $2.9 million per year. These sums do not provide for any increase in fleet capability.

Table 5 presents the result of the calculation, both in constant 1981 dollars and increased by an annual inflation rate of 7 percent. In the first year, the program would increase above the basic IDOE level by $8.9 million, the estimated additional cost of intermediate and pilot projects, ship refit and replacement, equipment replacement and development, and distant water research. Annual expenditures (1981 dollars) would increase to $57.9

TABLE 5 Estimated Costs of Post-IDOE Program[a]

	1981	1982	1983	1984	1985	1986	1987	1988	1989	1990
	In constant 1981 million dollars									
A	25.0	25.8	26.5	27.3	28.1	29.0	29.9	30.7	31.7	32.6
B	2.5	2.6	2.6	2.7	2.8	2.9	3.0	3.1	3.2	3.3
C	11.0	11.3	11.7	12.0	12.4	12.8	13.1	13.5	13.9	14.4
D	2.9	2.9	2.9	2.9	2.9	2.9	2.9	2.9	2.9	2.9
E	2.5	2.6	2.6	2.7	2.8	2.9	3.0	3.1	3.2	3.3
F	1.0	1.0	1.0	1.0	1.0	1.0	1.0	1.0	1.0	1.0
TOTAL	44.9	46.2	47.3	48.6	50.0	51.5	52.9	54.3	55.9	57.9
	In current million dollars (7 percent inflation rate)									
A	25.0	27.5	30.2	33.3	36.6	40.3	44.3	48.7	53.6	58.9
B	2.5	2.8	3.0	3.3	3.7	4.0	4.4	4.9	5.4	5.9
C	11.0	12.1	13.3	14.6	16.1	17.7	19.5	21.4	23.6	25.9
D	2.9	3.1	3.3	3.6	3.8	4.1	4.4	4.7	5.0	5.3
E	2.5	2.8	3.0	3.3	3.7	4.0	4.4	4.9	5.4	5.9
F	1.0	1.1	1.1	1.2	1.3	1.4	1.5	1.6	1.7	1.8
TOTAL	44.9	49.4	53.9	59.3	65.2	71.5	78.5	86.2	94.7	103.7

[a] A, Basic program; B, intermediate and pilot projects; C, ship operations; D, ship refit and replacement; E, equipment replacement and development; F, additional distant water research costs.

million (constant 1981 dollars) or $103.7 million (7 percent inflation rate) by 1990. This growth should also be reflected in support for individual investigators and in other appropriate aspects of the national oceanographic program.

The estimates do not provide for significant new scientific developments and opportunities nor for accelerated growth due to unanticipated social demands. Nonetheless, we consider them a reasonable basis for planning within the constraints of present knowledge.

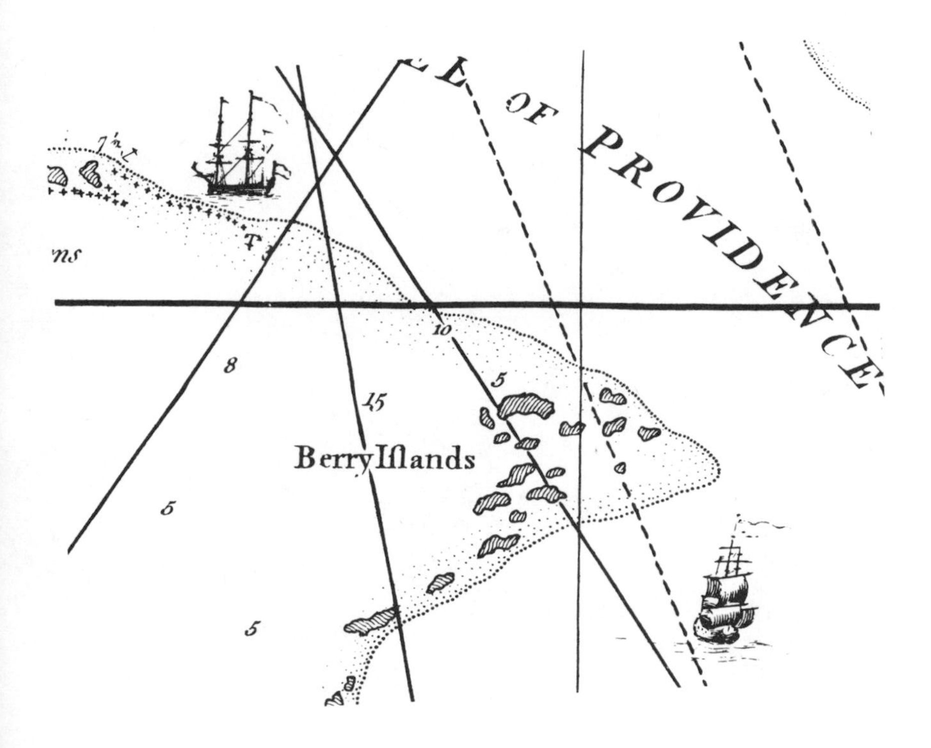

Appendix A
Acronyms and Abbreviations

CALCOFI	California Cooperative Oceanic Fisheries Investigations
CENOP	Cenozoic Paleo-oceanography Project
CEPEX	Controlled Ecosystem Pollution Experiment
CLIMAP	Climate Long Range Investigation Mapping and Prediction Study
COMS	Center for Ocean Management Studies
CTD	Conductivity, temperature, depth
CUEA	Coastal Upwelling Ecosystem Analysis
DSDP	Deep Sea Drilling Program
EASTROPAC	An Oceanographic Study of the Eastern Tropical Pacific
EDS	Environmental Data Service
EQUAPAC	An International Cooperative Survey of the Equatorial Pacific (1956)
FGGE	First GARP Global Experiment
GARP	Global Atmospheric Research Program
GATE	GARP Atlantic Tropical Experiment
GEOSECS	Geochemical Ocean Sections Study
IDOE	International Decade of Ocean Exploration
IOC	Intergovernmental Oceanographic Commission
IPOD	International Program of Ocean Drilling
ISOS	International Southern Ocean Studies

MODE/POLYMODE	Mid Ocean Dynamics Experiment/Joint U.S.-U.S.S.R. Mid Ocean Dynamics Experiment
NACOA	National Advisory Committee on Oceans and Atmosphere
NASA	National Aeronautics and Space Administration
NOAA	National Oceanographic and Atmospheric Administration
NODC	National Oceanographic Data Center
NORPAC	An International Cooperative Survey of the North Pacific (1955)
NORPAX	The North Pacific Experiment
NSF	National Science Foundation
ONR	Office of Naval Research
PRIMA	Pollution Response in Marine Animals
RMT	Rectangular midwater trawl
ROSCOP	Report of Observations/Samples Collected by Oceanographic Programs
SEAREX	Sea–Air Exchange Project
SEATAR	Southeast Asia Tectonics and Resources
SES	Seagrass Ecosystem Study
SOFAR	Sound Fixing and Ranging
UNCLOS	United Nations Conference on the Law of the Sea
UNOLS	University National Oceanographic Laboratory System
USGS	U.S. Geological Survey
XBT	Expendible bathythermograph
XSTD	Expendible salinity temperature depth meter

Appendix B
Participants, Observers, and Staff

POST-IDOE PLANNING WORKSHOP
SEATTLE, WASHINGTON
SEPTEMBER 7–13, 1977

BERNHARD ABRAHAMSSON, Denver Research Institute, University of Denver, Denver, Colorado

DAYTON ALVERSON, Northwest Fisheries Center, National Marine Fisheries Service, NOAA, Seattle, Washington

JOHN R. APEL, Pacific Marine Environmental Laboratory, NOAA, Seattle, Washington

GUSTAF O. S. ARRHENIUS, Scripps Institution of Oceanography, University of California, La Jolla, California

THOMAS S. AUSTIN, Environmental Data Service, NOAA, Washington, D.C.

D. JAMES BAKER, JR., Department of Oceanography, University of Washington, Seattle, Washington

ALBERT W. BALLY, Shell Oil Company, Houston, Texas

KARL BANSE, Department of Oceanography, University of Washington, Seattle, Washington

RICHARD T. BARBER, Marine Laboratory, Duke University, Beaufort, North Carolina

CHARLES C. BATES, U.S. Coast Guard, Washington, D.C.

ROBERT C. BEARDSLEY, Woods Hole Oceanographic Institution, Woods Hole, Massachusetts

VICTOR BOATWRIGHT, Electric Boat Division, General Dynamics, Groton, Connecticut

JACK W. BOLLER, Marine Board, National Research Council, Washington, D.C.

ALBERT L. BRIDGEWATER, Directorate for Astronomical, Atmospheric, Earth and Ocean Sciences, National Science Foundation, Washington, D.C.

WALLACE BROECKER, Lamont-Doherty Geological Observatory, Columbia University, Palisades, New York

DOUGLAS L. BROOKS, National Advisory Committee on Oceans and Atmosphere, Washington, D.C.

JOHN V. BYRNE, Oregon State University, Corvallis, Oregon

CLOY CAUSEY, Texaco, Inc., Bellaire, Texas

PRESTON CLOUD, Department of Geological Sciences, University of California, Santa Barbara, California

RITA R. COLWELL, Department of Microbiology, University of Maryland, College Park, Maryland

JOHN D. COSTLOW, JR., Marine Laboratory, Duke University, Beaufort, North Carolina

JAMES CURLIN, Congressional Research Service, Library of Congress, Washington, D.C.

JOHN M. EDMOND, Department of Earth and Planetary Sciences, Massachusetts Institute of Technology, Cambridge, Massachusetts

DAVID FLEMER, Ecological Effects Division, Environmental Protection Agency, Washington, D.C.

PAUL FYE, Woods Hole Oceanographic Institution, Woods Hole, Massachusetts

WILLIAM S. GAITHER, College of Marine Studies, University of Delaware, Newark, Delaware

ALAN J. GROBECKER, Division of Atmospheric Sciences, National Science Foundation, Washington, D.C.

M. GRANT GROSS, Chesapeake Bay Institute, Johns Hopkins University, Baltimore, Maryland

JOHN GULLAND, Aquatic Resources Survey and Evaluation Service, FAO, Rome, Italy

GORDON HAMILTON, Navy Oceanographic Research and Development Activity, Bay St. Louis, Missouri

G. ROSS HEATH, University of Rhode Island, Kingston, Rhode Island

G. HEMPEL, Institut für Meereskunde, Universität Kiel, Kiel, FRG

J. BRACKETT HERSEY, Office of Naval Research, Arlington, Virginia

CHARLES D. HOLLISTER, Department of Geology and Geophysics, Woods Hole Oceanographic Institution, Woods Hole, Massachusetts

FEENAN JENNINGS, Office for the IDOE, National Science Foundation, Washington, D.C.

MARY K. JOHRDE, Office for Oceanographic Facilities Support, Division of Ocean Sciences, National Science Foundation, Washington, D.C.

WILLIAM W. KELLOGG, National Center for Atmospheric Research, Boulder, Colorado

LAURISTON KING, Office of the IDOE, National Science Foundation, Washington, D.C.

JOHN A. KNAUSS, Provost for Marine Affairs, University of Rhode Island, Kingston, Rhode Island

KAZUO KOBAYASHI, (Observer), Ocean Research Institute, University of Tokyo, Tokyo, Japan

DALE C. KRAUSE, (Observer), Division of Marine Sciences, UNESCO, Paris, France

REUBEN LASKER, Southwest Fisheries Center, National Marine Fisheries Service, NOAA, La Jolla, California

BRIAN T. LEWIS, Department of Oceanography, University of Washington, Seattle, Washington

BRUCE T. MALFAIT, Office for the IDOE, National Science Foundation, Washington, D.C.

CLAYTON MCAULIFFE, Chevron Oil Field Research Co., La Habra, California

HELEN McCAMMON, Division of Biomedical and Environmental Research, Energy Research and Development Agency, Washington, D.C.

JAMES J. McCARTHY, Department of Biology, Harvard University, Cambridge, Massachusetts

THANE McCULLOH, U.S. Geological Survey, Seattle, Washington

DONALD L. McKERNAN, Marine Affairs Program, University of Washington, Seattle, Washington

FOSTER H. MIDDLETON, Department of Ocean Engineering, University of Rhode Island, Kingston, Rhode Island

J. ROBERT MOORE, Institute of Marine Science, University of Alaska, College, Alaska

THEODORE C. MOORE, JR., Graduate School of Oceanography, University of Rhode Island, Kingston, Rhode Island

JEAN CLAUDE MOURLON (Observer), Science Attache, French Embassy, Washington, D.C.

KENNETH H. NEALSON, Scripps Institution of Oceanography, University of California, San Diego, La Jolla, California

WILLIAM A. NIERENBERG, Scripps Institution of Oceanography, University of California, San Diego, La Jolla, California

JAMES O'BRIEN, Meteorology Annex, Florida State University, Tallahassee, Florida

NED A. OSTENSO, National Sea Grant Program, National Oceanic and Atmospheric Administration, Washington, D.C.

MAURICE RATTRAY, JR., Department of Oceanography, University of Washington, Seattle, Washington

ROGER REVELLE, Department of Political Science, University of California, San Diego, La Jolla, California

BRIAN J. ROTHSCHILD, Office of Energy and Strategic Research Planning, Department of Commerce, Washington, D.C.

JAMES M. SHARP, Gulf Universities Research Consortium, Houston, Texas

JOHN L. SHAW, Ocean Management, Inc., 300 120th Avenue, N.E., Bellevue, Washington

DEREK W. SPENCER, Department of Chemistry, Woods Hole Oceanographic Institution, Woods Hole, Massachusetts

R. W. STEWART, Marine Sciences Branch, Pacific Region, Department of the Environment, Victoria, B.C., Canada

HENRY STOMMEL, Woods Hole Oceanographic Institution, Woods Hole, Massachusetts

WILLIAM L. SULLIVAN, JR., Office of Marine Science and Technology Affairs, Department of State, Washington, D.C.

GEORGE H. SUTTON, Hawaii Institute of Geophysics, University of Hawaii, Honolulu, Hawaii

MANIK TALWANI, Lamont-Doherty Geological Observatory, Columbia University, Palisades, New York

CHRISTOPHER VANDERPOOL, Department of Sociology, Michigan State University, East Lansing, Michigan

JOHN M. WALLACE, Department of Atmospheric Sciences, University of Washington, Seattle, Washington

FERRIS WEBSTER, National Oceanographic and Atmospheric Administration, Rockville, Maryland

RAY F. WEISS, Scripps Institution of Oceanography, La Jolla, California

ELMER P. WHEATON, Vice-President and General Manager (ret.), Lockheed Missiles and Space Company, Portola Valley, California

WILLIAM F. WHITMORE, Ocean Systems, Lockheed Missiles and Space Company, Sunnyvale, California

FRANCIS WILLIAMS, Special Projects, NASA Headquarters, Washington, D.C.

WARREN S. WOOSTER, Institute for Marine Studies, University of Washington, Seattle, Washington

CARL WUNSCH, Department of Earth and Planetary Sciences, Massachusetts Institute of Technology, Cambridge, Massachusetts

Ocean Sciences Board Staff

RICHARD C. VETTER, *Executive Secretary*
JO ANN DYER,
ELLEN FUHRER